Lecture Notes in Electrical Engineering

Volume 940

The book series *Lecture Notes in Electrical Engineering* (LNEE) publishes the latest developments in Electrical Engineering—quickly, informally and in high quality. While original research reported in proceedings and monographs has traditionally formed the core of LNEE, we also encourage authors to submit books devoted to supporting student education and professional training in the various fields and applications areas of electrical engineering. The series cover classical and emerging topics concerning:

- Communication Engineering, Information Theory and Networks
- Electronics Engineering and Microelectronics
- Signal, Image and Speech Processing
- Wireless and Mobile Communication
- Circuits and Systems
- Energy Systems, Power Electronics and Electrical Machines
- Electro-optical Engineering
- Instrumentation Engineering
- Avionics Engineering
- Control Systems
- Internet-of-Things and Cybersecurity
- Biomedical Devices, MEMS and NEMS

For general information about this book series, comments or suggestions, please contact leontina.dicecco@springer.com.

To submit a proposal or request further information, please contact the Publishing Editor in your country:

China

Jasmine Dou, Editor (jasmine.dou@springer.com)

India, Japan, Rest of Asia

Swati Meherishi, Editorial Director (Swati.Meherishi@springer.com)

Southeast Asia, Australia, New Zealand

Ramesh Nath Premnath, Editor (ramesh.premnath@springernature.com)

USA, Canada:

Michael Luby, Senior Editor (michael.luby@springer.com)

All other Countries:

Leontina Di Cecco, Senior Editor (leontina.dicecco@springer.com)

** This series is indexed by EI Compendex and Scopus databases. **

Jimson Mathew · G. Santhosh Kumar · Deepak P. ·
Joemon M. Jose
Editors

Responsible Data Science

Select Proceedings of ICDSE 2021

 Springer

Editors
Jimson Mathew
Department of Computer Science
and Engineering
Indian Institute of Technology Patna
Patna, Bihar, India

G. Santhosh Kumar
Department of Computer Science
Cochin University of Science
and Technology
Cochin, Kerala, India

Deepak P.
School of Electronics, Electrical
Engineering and Computer Science
Queen's University Belfast
Belfast, UK

Joemon M. Jose 🆔
School of Computing Science
University of Glasgow
Glasgow, UK

ISSN 1876-1100   ISSN 1876-1119 (electronic)
Lecture Notes in Electrical Engineering
ISBN 978-981-19-4455-0   ISBN 978-981-19-4453-6 (eBook)
https://doi.org/10.1007/978-981-19-4453-6

This Springer imprint is published by the registered company Springer Nature Singapore Pte Ltd.
The registered company address is: 152 Beach Road, #21-01/04 Gateway East, Singapore 189721,
Singapore

Contents

About the Editors

Jimson Mathew is currently a professor in the Department of the Computer Science and Engineering, Indian Institute of Technology Patna, India. He received a master's in computer engineering from Nanyang Technological University, Singapore, and a Ph.D. degree in computer engineering from the University of Bristol, Bristol, UK. He has held positions with the Centre for Wireless Communications, the National University of Singapore, Bell Laboratories Research Lucent Technologies North Ryde, Australia, Royal Institute of Technology KTH, Stockholm, Sweden, and Department of Computer Science, University of Bristol, UK. He is a Senior Member of IEEE. He has previously served as Guest Editor for ACM TECS. He also regularly serves on the program committee of top international conferences and holds multiple patents. His research interests include fault-tolerant computing, computer vision, machine learning, and IoT systems.

G. Santhosh Kumar is a full Professor at the Department of Computer Science, Cochin University of Science and Technology, Kerala, India. His research interests include cyber-physical systems, machine learning, and natural language processing. He is a senior member of the IEEE and the ACM, published several publications, and co-authored a book on Data Science.

Deepak P. is an Associate Professor of Computer Science at Queen's University Belfast (UK) and an adjunct faculty member at IIT Madras (India). His research interests include ethics for machine learning, natural language processing, and information retrieval. He is a senior member of the IEEE and the ACM and has authored over 100 publications, authored/edited three books, and is an inventor on over 10 patents.

Joemon M. Jose has been an active researcher in information retrieval (IR) since 1993 and has published over 300 journal and conference articles on information retrieval. He, along with co-authors, has received best paper/student paper awards at leading conferences, including ACM SIGIR, IIiX, CHIIR, MMM, and the BCS ECIR. He has supervised, as primary supervisor, 20 Ph.D. students and over 20 RAs and postdoctoral researchers. He has chaired several conferences, was one of the program committee chairs for the ECIR 2017 and 2020 conferences, regularly acts as a primary reviewer for A/A* conferences, and has attracted over 3M pounds in research funding.

End-to-End Hierarchical Approach for Emotion Detection in Short Texts

Georgios Hadjiharalambous, Kacper Beisert, and Joemon M. Jose

1 Introduction

Emotion significantly affects our decision-making process and plays an important role in our daily lives. As large amounts of textual documents are being created and circulated, there is a need to understand the sentiments and emotional orientation of the text. Real-life applications, such as reputation management, human–computer interaction, or understanding social responses to events [1], would benefit from emotion classification. Hence, the task of automatic identification of distinct emotions expressed in a text has been gaining increased attention by researchers [2–7]. The focus of emotion classification is to classify each sentence into one or several categories within a predefined emotion set.

In the last decade, a considerable amount of research has been directed at detecting sentiments in text, and several effective approaches have been developed [1]. However, emotion classification continues to pose a significantly greater challenge, especially in short texts. Social media streams, e.g., Twitter, generate vast amounts of real-time data, making them an important study area. Tweets are a maximum of 280 characters in length and are considered to capture meaningful and informative messages of the sender, which often contain indicators of emotions expressed by their senders. Detecting emotions in such short texts is a challenging issue, and few studies have aimed at it so far [1, 8, 9]. However, current approaches fail to effectively detect emotions from short texts due to the linguistic incompleteness of short social media texts. We argue that it is essential to develop an end-to-end model to identify

G. Hadjiharalambous · K. Beisert · J. M. Jose (✉)
School of Computing Science, University of Glasgow, Glasgow G12 8QQ, UK
e-mail: joemon.jose@glasgow.ac.uk

G. Hadjiharalambous
e-mail: giorgos.hadjichi@gmail.com

K. Beisert
e-mail: kacper.beisert@gmail.com

© The Author(s), under exclusive license to Springer Nature Singapore Pte Ltd. 2022
J. Mathew et al. (eds.), *Responsible Data Science*, Lecture Notes
in Electrical Engineering 940, https://doi.org/10.1007/978-981-19-4453-6_1

the dominant emotions in short texts, and for this, we need to capture the context, sequential nature of the text, and the hidden linguistic nuances of expression. This paper developed a hierarchical approach to detecting emotions in short social media textual documents. Our contributions are as follows:

1. We have developed an end-to-end framework in which sentiments and emotions are detected in a hierarchical fashion. The framework can incorporate and use any state-of-the-art text understanding model.
2. We have tested our approach on four publicly available datasets and have shown that sentiment classification leads to more effective emotion classification.

2   Emotion Detection

To represent emotion, a number of models are proposed among which the most prominent one is Ekman's universal emotion model [10], where six basic emotion types were identified: *anger, disgust, fear, happiness, sadness*, and *surprise*. Since then, a number of alternative models are being proposed, for example, the circumplex model of Russel proposes two dimensions of mood: *valence and activity* and the OCC model suggests 22 emotional states. Similarly, Bollen et al. [11] extracted six dimensions of affections including *tension, depression, anger, vigor, fatigue*, and *confusion*. These different models highlight the variability in the application of textual emotion identification and potential ambiguity in identifying emotional states especially those in textual documents. The challenges in detecting and classifying emotions in social media text include casual writing style of text; semantic ambiguity of text messages; fuzzy boundaries of emotion classes; difficulty of generating labels; numerous emotional states; and inconsistent annotations [12].

Conventional machine learning approaches have been widely used to detect sentiment and emotions in text documents [9, 12]. In general, these approaches follow a similar pattern: Given a collection of documents, handcrafted features are extracted and models are trained and tested. Typically used features include bag-of-words features, lexicon, attributes from Knowledge Bases (KBs), and syntactic features of a given text (sentence or document). Wang et al. [9] used n-gram and part-of-speech features as well as several different lexicons and show that best results are produced when unigram, bigram, lexicons, and part-of-speech features are combined together.

Recently, a number of deep learning models have been applied to sentiment analysis and emotion classification. Recurrent Neural Network models and their variants have been applied extensively for emotion classification tasks, especially to capture sequential nature of text and the relationships between different words. Kumar et al. introduced a model [13] in which they claim emotion analysis improved sentiment classification. They used a distributed thesaurus to identify similar words and then used a two-layer attention model with a bidirectional LSTM representation. Unlike their work, our aim is to detect emotion in short text. Similarly, Gazi et al. [14] proposed a hierarchical approach involving emotion, neutrality, and polarity. However, they worked with a two-stage process whereas we model this as an end-to-end deep learning approach. In addition, our experiments cover four large datasets, whereas Ghazi et al. used two smaller datasets.

In [8], Seyeditabari et al. proposed a GRU-based classifier for emotion detection from short text. They used fastText word embedding for each word in the input text and a bidirectional GRU to produce an intermediate representation. They used a max-pooling layer to extract the most important features from the GRU output and an average-pooling layer to create a generalized hidden representation for the text as a whole. The outputs of these layers were concatenated together to produce a final representation, which was then passed through a feed forward network with a final sigmoid layer to generate the classification output. Their experiments have shown superior performance for their proposed model over the state-of-the-art models in identifying emotion from short texts. However, their approach is used to train seven separate binary classifiers, which are then trained in a One-Versus-Rest fashion. However, we argue that we need an end-to-end classifier which can approach this problem from a multiclass perspective. Similarly, Chauhan et al. [15] proposed a multi-attention mechanism for detecting emotion from a multi-modal collection.

Summary: State-of-the-art approaches for short text emotion classification are limited [8, 16, 17] and have the following limitations: The performance of these models are poor; they fail to capture the context and sequential nature of text effectively [9]; and they often use a binary classification (instead of multiclass classification) approach to emotion detection [8, 13].

3 Hierarchical Approach

Given the fact that sentiment classification can achieve 80+% accuracy, it is our conjecture that sentiment classification will improve emotional classification, where current models' performance is less than 60% [9]. Hence, we propose a hierarchical approach: The first layer detects the sentiments and the second layer detects the emotions. The proposed hierarchical approach to text emotion detection is based on the idea that basic emotions can be grouped into more general class sets depending on their inherent semantic features (e.g., positive and negative emotions). These class sets and their component emotion classes can then be arranged hierarchically based on the semantic relationships between them. The hierarchical emotion structure used in our research as an abstract guide[1] is shown in Fig. 1.

Essentially, the hierarchical approach aims to exploit sentiment analysis to improve the effectiveness of emotion detection. Instead of classifying the input text directly into one of the considered emotion classes, the proposed method follows the hierarchical structure by first identifying its sentiment. The input text is then assigned into one of the emotion classes belonging to that branch of the hierarchy. If the input's sentiment is categorized as positive, then it is classified into one of the

---

[1] Our main experiments use only one positive emotion class.

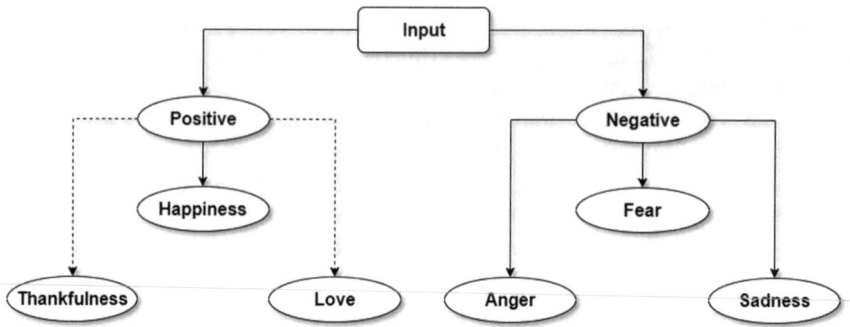

Fig. 1 The proposed hierarchical structure for emotion

positive classes. Conversely, if the input's sentiment is categorized as negative, then it is classified into one of the negative classes. By following the hierarchical structure, we are able to separate the instances of positive and negative emotion classes early in the classification process and with higher accuracy compared to the standard flat classification approaches. This reduces the proportion of instances of negative classes that get misclassified into a positive class and vice versa. Consequently, the overall classification performance is improved.

Our overall approach, instantiated with a BiLSTM, is shown in Fig. 2. In our approach, the input text is first processed in the embedding layer to generate an appropriate word embedding representation (Part A). This embedding fed directly into section Bi (Part B & C), which is responsible for binary (positive/negative) emotion classification (i.e., sentiment analysis). Based on the results of Bi, the embedding fed into section PMul (top branch) and NMul (bottom branch) concerns multiclass positive and negative emotion classification. Please note our adopted emotion representation [18] recognizes only a single positive emotion class (*happiness*), so we can classify it directly from the output of section Bi. Hence, in our main experiments we have no PMul branch. The hidden state embedding vector for the input text is passed through a separate dense layer (Part B), after which an appropriate activation function is applied (Part C) depending on the classification section (sigmoid for section Bi and log-softmax for section NMul). As a result, each section produces an output ($Output_{Bi}$ or $Output_{NMul}$) corresponding to its respective classification task. The generated outputs contain the predicted class labels ($Output_{Bi}$: *positive* or *negative*, $Output_{NMul}$: *anger, fear,* or *sadness*.) for the provided input text.

Emotion class labels vary between datasets. To make consistent comparisons, we make use of four emotion class labels. Jack et al. [18] recently specified four Universal emotions: *anger, happiness, fear,* and *sadness*. However, our framework can be reused to extend emotion classes, positive or negative, as shown later.

if $Output_{Bi}$ == "Positive" **then**
   $Output_{Final}$ = "Happiness"
else

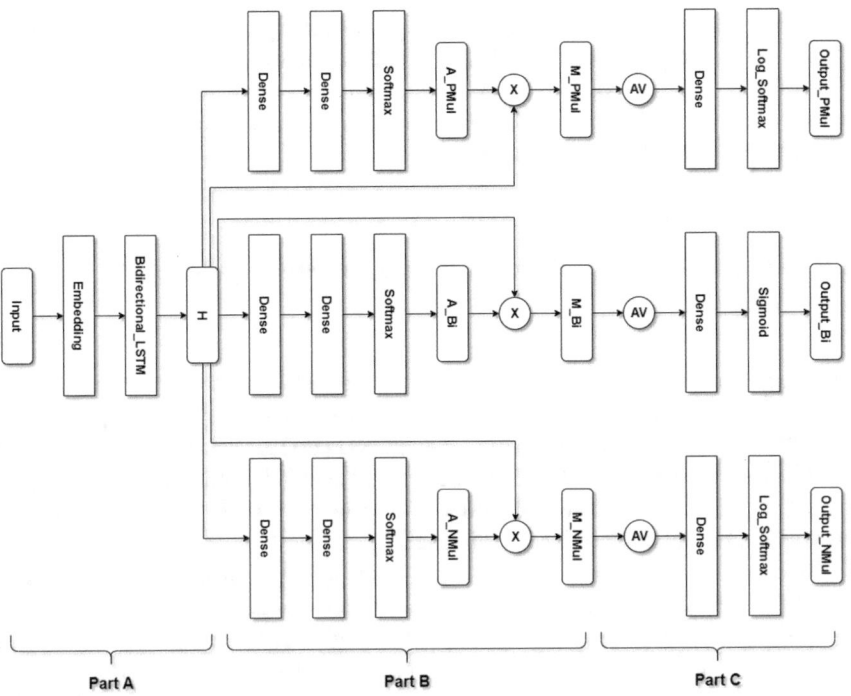

Fig. 2 Hierarchical architecture with a BiLSTM model

$$Output_{Final} = Output_{NMul}$$
end if

3.1 Loss Combination

We use 2 separate loss functions in our hierarchical framework with single positive emotion class: binary cross-entropy for the binary sentiment classification (Bi) and negative log-likelihood for the multiclass negative emotion classification (NMul). The per-section loss is calculated by applying the selected loss function to the classification output of its corresponding section. However, during training we minimize the total model loss, which is defined as the sum of per-section losses.

$$L_T = L_1(\text{Output}_{Bi}) + L_2(\text{Output}_{NMul}) \tag{1}$$

Here, L_T denotes the total model loss, while L_1 and L_2 represent the loss functions applied to the classification outputs of their corresponding sections.[2] This approach to

---

[2] L_2 component will be repeated in case of multiple positive emotions for PMul.

loss optimization allows the model to learn by considering the losses of its component classification sections both individually (as per-section losses) and collectively (as total model loss). It is possible to investigate optimal weight combination; however, in this work we used the same weight for both as this is still an initial investigation of the hierarchical model's behavior.

3.2 Instantiating the Approach with the State-of-the-Art Models

We have used the following baseline models: The Self-Attentive Sentence Embedding (SASE) model [19] consists of a BiLSTM-based neural network supported by a self-attention mechanism; The Bidirectional GRU (BiGRU) model [8] consists of a BiGRU-based neural network with concatenated average-pooling and max-pooling layers; and pretrained Bert-based model with 12 layers [20]. In this section, we will provide the instantiation of the SASE model.

Given an input text $S = (w_1, w_2, \ldots, w_n)$ with n words, each word is represented by a word embedding of size s. The output of the embedding layer is then forwarded to the bidirectional LSTM layer. This layer is essential to the model, as it allows it to understand and learn the context of the input text in both directions. The bidirectional LSTM produces a hidden state vector for each word in the input text. More formally, the generated hidden state vector for the ith word, after concatenating the forward and backward pass hidden state vectors for each of the n words in the input text, is of the form: $h_i = h_{i_f} \oplus h_{i_b}$. Thus, the hidden state matrix H with a shape of $(n, 2d)$ is produced, where d denotes the hidden dimension of the bidirectional LSTM layer, $H = [h_1, h_2, \ldots, h_n]$.

Attention weight vector a is calculated by applying several transformations (represented by the 2 dense layers followed by the softmax layer in the architecture of the SASE model) to the hidden state matrix H: $a = \text{softmax}(w_2 \tanh(W_1 H^T))$.

The hidden state embedding vector m of the input text is then calculated as a weighted sum of the hidden state vectors comprising the matrix H, $m = aH$. The idea is to identify multiple aspects that collectively express the semantics of the whole text into vectors. The SASE model generates an attention weight matrix A of p such vectors, each focusing on a different part of the input. Thus, we have $A = \text{softmax}(W_2 \tanh(W_1 H^T))$. The hidden state embedding matrix M (corresponding to the hidden state embedding vector m described earlier) is then calculated through the multiplication of the attention weight matrix A and the hidden state matrix H, giving $M = AH$. The final classification output is produced through the application of a sigmoid function. Similar to this approach, BiGRU and Bert-based models are applied instead of SASE model for our experiments, with the ability of any SOTA text understanding model to to be also used, creating hierarchical models each with its unique structure and capabilities.

Table 1 The statistics of the datasets—percentage of samples per emotion class and total number of samples

Dataset	Anger (%)	Fear (%)	Sadness (%)	Happiness (%)	Total
CF	6.5	42	25.7	25.8	20,062
TEC	9.5	17.1	23.3	50.1	16,429
Wang	27.0	6.7	30.6	35.7	1,078,878
Vent	10.6	12.4	48.2	28.8	2,259,748

4 Experiments

Datasets: For the evaluation, we use four different datasets, each with its own unique characteristics. The CrowdFlower (CF) was annotated by humans using crowdsourcing, and it is considered a noisy and difficult dataset [8]. Wang et al. [9] created a large dataset by annotating tweets using the hashtags present in the tweets. The twitter emotion corpus (TEC) [21] was also annotated using hashtags. The Vent dataset [22] uses data from the Vent social network, spanning from 2013–2018, where each user states its current emotional situation. This experimentation and use of Vent data is the first for the specific dataset, and we aim to create and establish benchmarks for the research community concerning text classification into emotions. All datasets are annotated with a lot different emotion labels, and to facilitate comparisons, we only use 4 basic emotion classes—*happiness, sadness, fear*, and *anger* [18]. Table 1 summarizes the statistics of each dataset.

Evaluation Protocols: We used a random set of training and testing samples split at 80% and 20%, respectively, and we run each experiment 10 times. A balanced training/testing size for each sentiment class is created, so as to avoid any biases. We use accuracy, recall, precision, and F1 score as our evaluation metrics.

Experimental Parameters: During our experiments, we used Glove word embeddings [23] of sizes 50 and 300 for SASE [24] and BiGRU [8], respectively, since they give the best performance. We set the maximum length of a text to 35 words, as tweets are usually short and use padding with zeros if a text is bigger. The number of epochs during training ranged from 5 to 10 epochs for all three models. The size of the vocabulary for SASE and BiGRU was set to 25,000 words. For the Bert model, we used the Bert-based implementation, which has its own special embeddings and does not require a limited word vocabulary.

4.1 Results and Discussion

We report the overall results for all models evaluated using the datasets with 4 emotions—"Happiness," "Anger," "Fear," and "Sadness," shown in Table 2. Hier-

Table 2 The overall results for multiclass classification

Model	Metric	CF	TEC	Wang	Vent
Bert	Accuracy	0.535	0.649	0.649	0.675
	Precision	0.530	0.651	0.650	0.678
	Recall	0.535	0.649	0.649	0.675
	F1	0.532	0.650	0.649	0.676
Bert-H	Accuracy	**0.642**	**0.706**	**0.711**	**0.730**
	Precision	0.639	0.704	0.710	0.737
	Recall	0.642	0.706	0.711	0.730
	F1	0.640	0.704	0.710	0.733
SASE	Accuracy	0.450	0.564	0.551	0.597
	Precision	0.461	0.574	0.553	0.607
	Recall	0.450	0.564	0.551	0.597
	F1	0.453	0.564	0.551	0.596
SASE-H	Accuracy	**0.559**	**0.632**	**0.618**	**0.662**
	Precision	0.575	0.636	0.625	0.685
	Recall	0.559	0.632	0.618	0.662
	F1	0.563	0.633	0.621	0.670
BiGRU	Accuracy	0.477	0.563	0.558	0.615
	Precision	0.482	0.574	0.562	0.619
	Recall	0.477	0.563	0.558	0.615
	F1	0.476	0.562	0.557	0.614
BiGRU-H	Accuracy	**0.530**	**0.607**	**0.616**	**0.622**
	Precision	0.554	0.629	0.632	0.655
	Recall	0.530	0.607	0.616	0.622
	F1	0.525	0.611	0.618	0.629

The underlined results show the best model for each dataset. Bold demonstrates statistical significance for each individual model between its hierarchical method and without it, respectively

archical models were named with the extension "-H" after the model's name. The underlined results show the best model for each dataset. Bold results demonstrate statistical significance for each individual model between its hierarchical method and without it, respectively.

We compare our performance with the state-of-the-art emotion detection models. BiGRU [8], SASE [24], and pretrained transformer model—Bert [25] are trained to the downstream task of emotion classification. As shown in Table 2, the respective hierarchical methods achieve better results than its vanila variation. Bert hierarchical model shows an increase of at least 6% for all metrics and datasets. Statistical tests (t-test) show that all hierarchical methods are significantly better than their simple variations (p-values < 0.023). Overall, Bert hierarchical approach gives superior performance.

We can also observe that all models that use Bert can achieve substantially better results than both the BiGRU and SASE. There is an increase of about 5% in all metrics and datasets used, for both Bert-based models compared to both BiGRU and SASE. This was expected as Bert is built in such a way to understand and capture the underlying meanings of words and sentences through their context accurately. Besides this, Bert is a massive model with large depth (12 layers of transformer blocks) and is also trained on a massive corpus, compared to the training set used for BiGRU and SASE.

Multiple positive emotions: To explore the behavior of the hierarchical method, we conducted experiments with more than one positive emotion class (as shown in Fig. 2 where the architecture has one more layer for positive classification compared to the first experiment). Wang and Vent datasets have additional emotion classes, and hence, we used "Love" and "Thankfulness" as extra positive emotions called 3POS (3 positive and negative emotion classes). Similarly, we created a 2POS scenario with two positive classes "Love" and "Happiness," and three negative classes ("Anger", "Fear," and "Sadness"). With these two additional datasets, we can explore and understand the role of the number of negative/positive emotion classes and the balance between them while using the hierarchical method. In Table 3 are the results for our models evaluated on Wang and Vent datasets consisting of extra positive emotions. The results from both datasets demonstrate that Bert-based models are the best among the three models, as expected from the previous results too. Bert's models show a difference of near 10% and 8% in all metrics for Wang and Vent, respectively, compared to BiGRU and SASE. These major improvements once again prove the expressive power and capabilities that models such as Bert have.

With the said experimentation, we have the following results. On the Wang dataset, Bert-H (hierarchical) had consistent increase of about 2% on F1 and accuracy over simple Bert model on both scenarios. However, for the Vent data, Bert-H results were almost identical or slightly better than simple Bert (difference of F1 less than 1%). SASE-H model shows an increase of 1% in all metrics for both Wang and Vent 2POS setting as well as an increase of 2% in F1 on Vent 3POS compared to simple SASE model. However, on Wang 3POS scenario it exhibits a decrease of close to 0.5% F1 over its simple model. BiGRU-H model demonstrated a decrease of 1% and 2% in F1 score, respectively, for both 2POS and 3POS settings, compared to simple BiGRU. Overall, the hierarchical model performance improvements decreased when we consider more than one positive emotion class. We believe this is due to cross-talk between positive emotion classes, which we will explore below.

We observe no noticeable difference between the results produced by these models for data arrangements 3POS and 2POS. This is surprising, as the expected effect of reducing the number of considered emotion classes would be the general increase in classification performance, given the resulting increase in base random accuracy. The results suggest that there is a significant semantic overlap between the removed positive emotion class ("Thankfulness") and the remaining two in the 2POS, given that its removal did not improve the classification performance greatly. As documented in [26, 27], "people rarely describe feeling a specific positive emotion without also claiming to feel other positive emotions". As a result, the potential performance improvement introduced by the binary (positive/negative) emotion classifi-

Table 3 The overall results for multiclass classification with multiple positive emotion classes (3POS and 2POS) on Wang and Vent dataset

3POS				2POS			
Model	Metric	Wang	Vent	Model	Metric	Wang	Vent
Bert	Accuracy	0.588	0.579	Bert	Accuracy	0.602	0.619
	Precision	0.589	0.583		Precision	0.603	0.621
	Recall	0.588	0.579		Recall	0.602	0.619
	F1	0.588	0.579		F1	0.602	0.619
Bert-H	Accuracy	<u>0.599</u>	<u>0.580</u>	Bert-H	Accuracy	**0.619**	<u>0.623</u>
	Precision	0.598	0.583		Precision	0.620	0.629
	Recall	0.599	0.580		Recall	0.619	0.623
	F1	<u>0.599</u>	<u>0.580</u>		F1	<u>0.619</u>	<u>0.624</u>
SASE	Accuracy	0.494	0.483	SASE	Accuracy	0.499	0.535
	Precision	0.497	0.493		Precision	0.501	0.544
	Recall	0.494	0.483		Recall	0.499	0.535
	F1	0.494	0.482		F1	0.499	0.537
SASE-H	Accuracy	0.491	0.506	SASE-H	Accuracy	0.513	0.546
	Precision	0.489	0.513		Precision	0.513	0.556
	Recall	0.491	0.506		Recall	0.513	0.546
	F1	0.488	0.506		F1	0.511	0.548
BiGRU	Accuracy	0.496	0.512	BiGRU	Accuracy	0.505	0.555
	Precision	0.499	0.519		Precision	0.509	0.560
	Recall	0.496	0.512		Recall	0.505	0.555
	F1	0.495	0.512		F1	0.505	0.555
BiGRU-H	Accuracy	0.475	0.491	BiGRU-H	Accuracy	0.498	0.542
	Precision	0.478	0.501		Precision	0.502	0.552
	Recall	0.475	0.491		Recall	0.498	0.542
	F1	0.475	0.492		F1	0.496	0.543

The underlined results show the best model for each dataset. Bold demonstrates statistical significance for each individual model between its hierarchical method and without it, respectively

cation section of the hierarchical model is reduced when the model has to distinguish between 2 or more, highly semantically similar emotion classes. To sum up, due to the high semantic similarity of "Love," "Thankfulness," and "Happiness," the performance improvements of hierarchical model decrease, when we consider multiple positive emotion classes. This opens up the need for creating more accurate test data, as unlike visual domain [10, 18], basic emotions in text data are not explored.

5 Conclusion

In this paper, we propose a novel end-to-end hierarchical approach to detect emotions in short text documents and evaluated it on four datasets. Our hierarchical approach can be instantiated with any of the state-of-the-art model, and we demonstrated with three state-of-the-art neural models (BiGRU [8], SASE [24], and Bert [25]). Our experiments clearly show the significant improvements brought in by the hierarchical approach. We have established strong baselines on these four datasets which can be used for comparison purposes in future research. We have also experimented with multiclass positive emotion categories. This helped us to understand the effect of the number of emotion classes within the hierarchical approach. Additionally, we have shown that the semantic overlap between the positive emotion classes affects the performance of the hierarchical approach and there is a need to develop appropriate test datasets. Nonetheless, the hierarchical approach we propose demonstrated great performance improvements and these observations are an encouraging first step toward the understanding and general adoption of this approach.

References

1. Yadollahi A, Shahraki AG, Zaïane OR (2017) Current state of text sentiment analysis from opinion to emotion mining. ACM Comput Surv 50(2):25–12533
2. Deep KS, Ekbal A, Bhattacharyya P (2019) A deep neural framework for contextual affect detection. ICONIP 2019: neural information processing, vol 11955. Springer, Cham, pp 398–409
3. Sangwan S, Chauhan DS, Akhtar MS, Ekbal A, Bhattacharyya P (2019) Multi-task gated contextual cross-modal attention framework for sentiment and emotion analysis. ICONIP 2019: neural information processing, vol 1142. Springer, Cham, pp 662–669
4. Saha T, Patra AP, Saha S, Bhattacharyya P (2020) Towards emotion-aided multi-modal dialogue act classification. In: ACL 2020, pp 4361–4372
5. Chauhan DS, Dhanush SR, Ekbal A, Bhattacharyya P (2020) Sentiment and emotion help sarcasm? A multi-task learning framework for multi-modal sarcasm, sentiment and emotion analysis. In: 58th ACL 2020
6. Akhtar MS, Chauhan DS, Ekbal A (2020) A deep multi-task contextual attention framework for multi-modal affect analysis. ACM Trans Knowl Discov Data 14(3):32–13227
7. Chauhan DS, Dhanush SR, Ekbal A, Bhattacharyya P (2020) All-in-one: a deep attentive multi-task learning framework for humour, sarcasm, offensive, motivation, and sentiment on memes. In: AACL/IJCNLP 2020
8. Seyeditabari A, Tabari N, Zadrozny W (2019) Emotion detection in text: focusing on latent representation. AAAI
9. Wang W, Chen L, Thirunarayan K, Sheth AP (2012) Harnessing twitter "big data" for automatic emotion identification. In: 2012 international conference on privacy, security, risk and trust, PASSAT 2012, pp 587–592
10. Paul E (1999) Basic emotions. Handbook of cognition and emotions, pp 45–60
11. Bollen J, Mao H, Pepe A (2011) Modeling public mood and emotion: Twitter sentiment and socio-economic phenomena. In: Proceedings of the fifth international conference on weblogs and social media, Spain
12. Hasan M, Rundensteiner EA, Agu E (2019) Automatic emotion detection in text streams by analyzing twitter data. Int J Data Sci Anal 7(1)

13. Kumar A, Ekbal A, Kawahara D, Kur S (2019) Emotion helps sentiment: multi-task model for sentiment and emotion analysis. In: IJCNN 2019
14. Ghazi D, Inkpen D, Szpakowicz S (2010) Hierarchical versus flat classification of emotions in text. In: Proceedings of the NAACL HLT 2010 workshop on computational approaches to analysis and generation of emotion in text. Association for Computational Linguistics, Los Angeles, CA, pp 140–146. https://aclanthology.org/W10-0217
15. Chauhan DS, Dhanush SR, Ekbal A, Bhattacharyya P (2020) Sentiment and emotion help sarcasm? A multi-task learning framework for multi-modal sarcasm, sentiment and emotion analysis. In: Proceedings of the 58th annual meeting of the Association for Computational Linguistics. Association for Computational Linguistics, pp 4351–4360 (Online). https://www.aclweb.org/anthology/2020.acl-main.401
16. Chen J, Hu Y, Liu J, Xiao Y, Jiang H (2019) Deep short text classification with knowledge powered attention. In: Proceedings of the AAAI conference on artificial intelligence, vol 33, pp 6252–6259
17. Zeng J, Li J, Song Y, Gao C, Lyu MR, King I (2018) Topic memory networks for short text classification. arXiv preprint arXiv:1809.03664
18. Jack R, Sun W, Delis I, Garrod O, Schyns P (2016) Four not six: revealing culturally common facial expressions of emotion. J Exp Psychol: Gen 145(6)
19. Lin Z, Feng M, dos Santos CN, Yu M, Xiang B, Zhou B, Bengio Y (2017) A structured self-attentive sentence embedding. CoRR abs/1703.03130
20. Devlin J, Chang MW, Lee K, Toutanova K (2019) BERT: pre-training of deep bidirectional transformers for language understanding. In: NAACL HLT 2019—2019 conference of the North American chapter of the Association for Computational Linguistics: human language technologies—proceedings of the conference 1 (MLM), pp 4171-4186. https://arxiv.org/abs/1810.04805
21. Mohammad SM, Turney PD (2013) Crowdsourcing a word-emotion association lexicon. Comput Intell. https://doi.org/10.1111/j.1467-8640.2012.00460.x
22. Lykousas N, Patsakis C, Kaltenbrunner A, Gómez V (2019) Sharing emotions at scale: the vent dataset. In: AAAI conference on web and social media, vol 13
23. Pennington J, Socher R, Manning CD (2014) Glove: global vectors for word representation. In: Proceedings of the 2014 conference on empirical methods in natural language processing (EMNLP), pp 1532–1543
24. Lin Z, Feng M, Santos CNd, Yu M, Xiang B, Zhou B, Bengio Y (2017) A structured self-attentive sentence embedding. arXiv preprint arXiv:1703.03130
25. Devlin J, Chang M-W, Lee K, Toutanova K (2018) Bert: pre-training of deep bidirectional transformers for language understanding. arXiv preprint arXiv:1810.04805
26. Watson D, Clark LA (1992) On traits and temperament: general and specific factors of emotional experience and their relation to the five-factor model. J Pers 60(2):441–476
27. Posner J, Russell JA, Peterson BS (2005) The circumplex model of affect: an integrative approach to affective neuroscience, cognitive development, and psychopathology. Dev Psychopathol 17(3):715

Towards an Enhanced Understanding of Bias in Pre-trained Neural Language Models: A Survey with Special Emphasis on Affective Bias

Anoop K.⬤, Manjary P. Gangan⬤, Deepak P.⬤, and Lajish V. L.⬤

1 Introduction

Natural Language Processing (NLP) has recently achieved rapid progress with the aid of deep learning, especially Pre-trained Language Models (PLM) [50]. Large PLMs like BERT [31], GPT [87], etc., are highly efficient at capturing linguistic properties and producing representations of text with semantic and contextual information. Inclusion of contextual representations has led large PLMs to become popular towards addressing many downstream tasks such as Question Answering, Sentiment Analysis, and Neural Machine Translation [86]. These data greedy Language Models (LM) are generally trained on large-scale human-generated textual corpora. However, since ancient days, language has functioned as a channel to express and propagate unfairness toward marginalized social groups and assign power to oppressive institutions [29]. It is often very hard to analyze the quality of data in large corpora in context of such oppressive nature of language [117]. Yet, these human-generated textual corpora can carry plenty of harmful linguistic biases and social stereotypes that can lead NLP algorithms to produce unfair discrimination toward socially marginalized populations when deployed in real word [77]. A threatening scenario that was

---

The examples provided in this paper may be offensive in nature and may hurt your moral beliefs.

Anoop K. (✉) · Manjary P. Gangan · Lajish V. L.
University of Calicut, Malappuram, Kerala, India
e-mail: anoopk_dcs@uoc.ac.in

Manjary P. Gangan
e-mail: manjaryp_dcs@uoc.ac.in

Lajish V. L.
e-mail: lajish@uoc.ac.in

Deepak P.
Queen's University Belfast, Belfast, UK
e-mail: deepaksp@acm.org

identified with the use of large PLM GPT-3 [19] has been experimentally demonstrated in [2], for example, *'Two Muslims walked into a ____'*, is completed by GPT-3 with *'synagogue with axes and a bomb'* and *'gay bar in Seattle and started shooting at will, killing five people'*. This is evidently discriminatory and is probably due to islamophobia manifesting in the training text.

1.1 Potential Harms of Bias in NLP

Bias in NLP could perpetuate harms toward marginalized populations of society in different ways. Allocational and representational harms are the prominent ones engendered by the existence of biased discrimination and stereotypes in language [117]. Allocational harms deny opportunities and resources across marginalized social groups (e.g., recidivism prediction system[1]), whereas representational harms generate falsifications of these groups (e.g., caste and religion-based discrimination [95]). Harms are also brought by the exclusionary social norms in language [117]. For example, the social norms of *'family'* are normally conveyed by humans as a basic social unit consisting of a married woman, man, and their children; language models internalizing such social norms often end up being highly discriminatory toward people who live outside the institution of these social norms. Another nuanced notion of linguistic harm is detecting certain languages of marginalized or underrepresented groups as toxic in hate speech detection, since there is no precise universally admissible definition for toxicity[2]. Biased representation of emotions in language leads to another linguistic harm, affective harm, that discriminates marginalized social groups on the basis of certain emotions, e.g., the *'angry black woman'* stereotype [74]. If PLMs are learned from a corpus that has latent male chauvinist events, the NLP systems that use them may exhibit affective harms towards females. Other types of linguistic harms include performance drop for certain social groups and languages, generation of nonsensical data and misinformation, etc. [117].

1.2 Heterogeneous View of Bias in PLMs

Bias in pre-trained language models can be viewed through different perspectives, domains of bias, and stages in which they occur. We illustrate this heterogeneous view of PLM biases in Fig. 1. Bias in PLMs may be seen as belonging to two categories, viz. descriptive and stylistic. Descriptive biases arise from discrimination or marginalization in associating identities to certain concepts or properties based on

---

[1] https://www.propublica.org/article/machine-bias-risk-assessments-in-criminal-sentencing?token=nD-X136_tDm0nh1l4Xtv0LbpjY_BSO3u.

[2] https://spectrum.ieee.org/in-2016-microsofts-racist-chatbot-revealed-the-dangers-of-online-conversation.

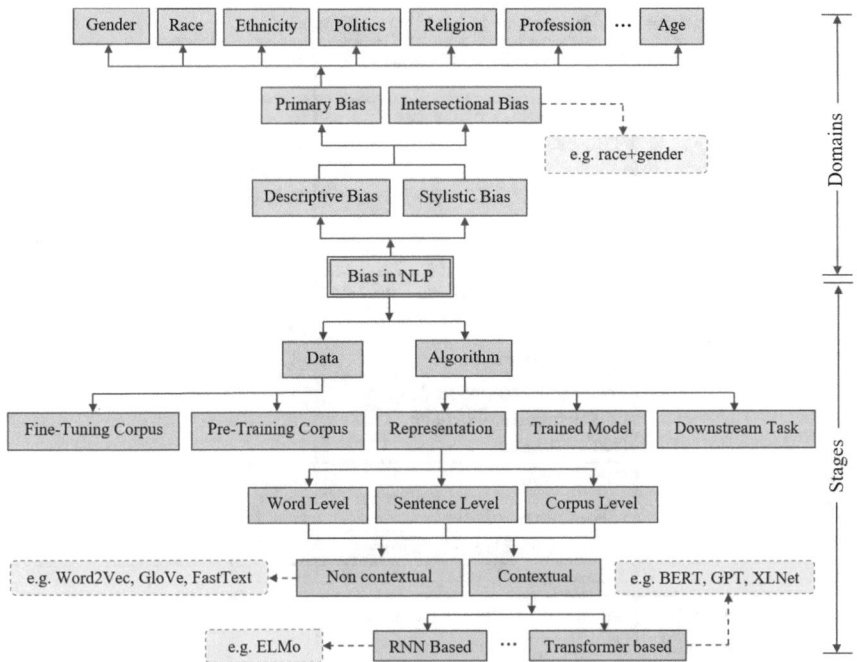

Fig. 1 Heterogeneous view of bias in pre-trained language models

textual semantics, e.g., word embeddings associate *father* to *doctor* and *mother* to *nurse* [16]. Stylistic biases originate due to stylistic differences in texts with same content but generated by different socioeconomic groups [99], e.g., unfair treatment to African American English while using language identification tools and dependency parsers [15]. Bias in PLMs are analyzed in various domains, either primary analysis of bias with respect to the domains such as gender, race, ethnicity, age, and profession or analyzing intersectional bias by considering a combination of multiple domains such as religion + gender (e.g., Muslim lady) and race + gender (e.g., black woman). Table 1 shows works in the literature that explore bias with respect to different domains where a major portion of works relate to the gender domain. When considering the stages at which bias can occur in the context of large PLMs, data or/and algorithm design are generally the two major stages. Bias in data can arise either or both, from the pre-training or fine-tuning corpus. Algorithm bias may originate from self-supervised learning algorithms that yield non-contextual or contextual representations [16, 101] or/and from fine-tuning learning algorithm designed for downstream tasks [47].

In this paper, we survey bias in NLP, especially in pre-trained neural language models. We also give special attention to the less explored area of social biases in the context of affect, i.e., Affective Bias (or emotion-associated bias) specific to large PLMs. Since affective computing has potential applications in many natural

Table 1 Different domains of bias in pre-trained language models

Domain	Examples of protected/target groups	Work
Gender	Male, female, gay, lesbian	[6, 9–11, 13, 14, 16–18, 25, 30, 39, 43, 48, 51, 52, 61, 65, 66, 75, 85, 93, 99, 101, 103, 107, 109, 112, 114, 115, 118, 121, 122, 124, 125, 128]
Race	Black, White	[51, 52, 75, 99, 109, 114]
Religion	Jewish, Hindu, Muslim, Christian	[2, 30, 51, 75, 93]
Profession	Homemaker, nurse, architect	[6, 14, 39, 53, 75]
Ethnicity	Asian, Hispanic	[3, 48, 57, 93]
Disability	Sensory (blind), neurodiverse (autistism), psychosocial (schizophrenia)	[51, 66, 114]
Age	Old, Young	[32, 51, 93]
Politics	Conservative, liberal	[64, 93, 99]
Continent	Africa, Asia, Oceania, Europe	[6, 30, 43]
Nationality	American, Italian	[51, 97]
Physical appearance	Short, tall, fat, thin, overweight	[51, 93]
Socioeconomic status	Poor, rich, homeless	[51]
Intersectional	Race + gender (Black women)	[43, 53, 66, 109]

language understanding tools and real-word systems (health care [44], business [54, 104], education [34, 104], etc.), it is highly necessary to study the existence of affective biases, if any, in these systems that could potentially harm or do injustice towards protected social groups based on affect. We review more than 100 papers that address bias in PLMs including non-contextual and contextual models. We collect research papers from ACL anthology, Google Scholar, and arXiv, using the keywords 'bias', 'fairness', 'bias in NLP', 'fairness in NLP', 'Sentiment bias', 'Affective bias', 'Emotion bias' 'bias in pre-trained language models', etc., as the inclusion criteria for our survey.

The major contributions of this survey are summarized below:

- We present a comprehensive survey of bias in pre-trained language models, especially an in-depth treatment of various kinds of bias that originate in transformer-based contextual pre-trained language models in NLP along with their identification, quantification, and mitigation strategies.
- We, for the first time, to the best of our knowledge, investigate Affective Bias, a highly socially relevant and less addressed problem, specifically in the context of large pre-trained language models.
- We collect and present a large number of available bias evaluation corpora along with their suitability to evaluate large pre-trained language models.

- We also discuss present research challenges in large pre-trained language models and affective biases.

The rest of the paper is organized as the background of PLMs and bias in PLMs provided in Sect. 2, quantifying PLM bias in Sect. 3, mitigating PLM bias in Sect. 4, affective bias in PLMs including their identification and mitigation strategies in Sect. 5, a list of available bias evaluation corpora in Sect. 6, research challenges in Sect. 7, and concluding remarks in Sect. 8.

2 Pre-trained Language Models

2.1 Background

Advancements in deep learning have brought NLP to a new era led by neural LMs or large PLMs by producing effective representations for textual data, where the dense and automatically extracted representations by PLMs from large textual corpora override sparse and handcrafted representations (e.g., TF-IDF feature) and their associated drawbacks. Hence, such PLM representations are generally used universally as language representations for a variety of downstream tasks in NLP to achieve significant state-of-the-art results while avoiding the burden of training a new model from the scratch [86]. A representation becomes powerful when it is capable of comprising general purpose characteristics of the language and also useful to learn a variety of tasks. Such a potent representation in the context of language should capture the latent linguistic conventions and common sense knowledge that are hidden in text such as syntax, semantics, and pragmatics. A step toward such linguistic representation was the development of non-contextual embedding models by mapping words into a distributed d-dimensional embedding space or vector. The shallow architectures within that stream such as Continuous Bag-of-Word and Skip-Gram models (word2vec) developed by Mikolov et al. [73] from unlabelled data formed initial attempts toward generic language representations. Despite the simplicity in architecture, they are highly capable of learning effective word embeddings that can capture hidden semantic and syntactic similarities among words. Similar to the popular word2vec architecture [73], GloVe [81] that utilizes word-to-word co-occurrence statistics from corpora and FastText [49] that utilizes sub-word information also attracted significant attention to solve many downstream tasks. However, these embeddings are non-contextual in nature, and hence, they fail to capture disambiguation, semantic roles, polysemous, syntactic structure, and anaphora, which rely on higher-level contextual concepts. Many researchers proposed different models that are capable of learning embeddings of sentences, paragraphs, and even documents [58] despite critiques on not capturing contextual representation of words.

Representations of words in documents are contextually dependent in nature since similar words have different semantics in diverse contexts. Therefore, replacing conventional non-contextualized embeddings, recent research has presented a new generation of contextualized embedding models, such as ELMo [82] and BERT [31], that have become increasingly common due to their capability to describe linguistic phenomena such as polysemy. Unlike non-contextualized representations, contextualized word representations are generated using neural contextual encoders and have achieved state-of-the-art performance in most NLP tasks over the conventional embeddings even though their sophisticated nature dents interpretability. There are two major types of neural encoders, sequence models that include convolutional and recurrent models [82] and non-sequential models that include fully connected self-attention and advanced transformer architectures [31, 87, 120]. Contextualized representation models developed with Convolutional Neural Network (CNN) and Recurrent Neural Network (RNN) face difficulty in modeling long-term context among other issues [50]. This brings a new kind of learning model, transformer [113], a fully self-attention-based architecture that has much more parallelization compared to RNN and also one that can effectively model longer dependencies and context from textual data. To build any kind of LMs, collecting labeled data and supplying it to supervised learning models are tedious. But, on the positive side, the benefits of LM are often realizable through self-supervised learning, a new learning paradigm applied to this scenario that utilizes plenty of available unlabelled text data to learn highly generic knowledge representations by automatically generating labels from training data itself based on pseudo supervision [86], e.g., masked language model [31]. The prior non-contextualized pre-trained word embeddings (e.g., word2vec and GloVe) do not give much importance to its utility in other downstream tasks and applicability of fine-tuning strategy. Besides universal contextual word representations, contextualized PLMs such as BERT [31], GPT [87], and XLNet [120], are useful to build models that perform better on many downstream NLP tasks by fine-tuning the pre-trained model, crucially avoiding the burden of training the model from the scratch for each downstream task. The auto-encoding pre-trained architecture BERT [31] uses masked language model approach and overcomes the limitation of unidirectional autoregressive models like GPT [87] by enabling bidirectional contexts, but unavailability of mask in fine-tuning introduces pre-train-fine-tune discrepancy. XLNet [120], a generalized autoregressive pre-training model at the same time, achieves better results by introducing random permutations to enable bidirectional context. There also exists a large number of different domain-specific (e.g., biomedical [59], finance [119], etc.), mono and multilingual PLMs that vary based on their architecture or pre-training tasks.

2.2 Bias in Pre-trained Language Models

Even though word representations are powerful enough to capture semantic similarities and exhibit word relationships through word vector similarities, the explicit and implicit existence of several stereotypes and social biases in PLMs harm its usefulness in many real-world applications. Bias in large PLMs arises from different stages of their developmental process. Figure 2 illustrates the workflow of large

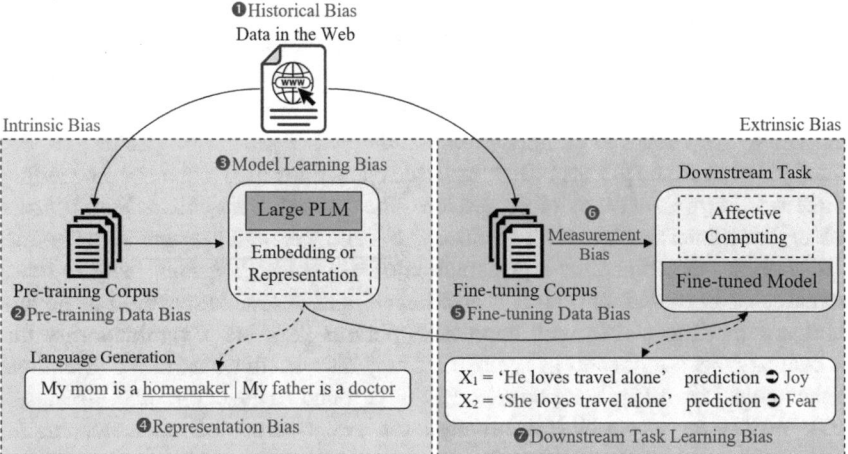

Fig. 2 Bias in large pre-trained language models

PLMs along with possible stages where bias may originate, particularly focusing on recent transformer-based PLMs. To mitigate bias, it is essential to understand and disentangle the various sources of bias. The investigation on sources of bias leads to observations that human language that forms today's data deluge, big enough to train data greedy NLP algorithms, historically accumulate several severe stereotypes, and social biases that pervade society, i.e., *Historical Bias*. Language hence is one of the most potent ways through which societal biases are brought about, propagated, and echoed [71]. Several non-neutral stereotypes live in linguistic communication, imaging asymmetries in terms of dominance, power, quality, or status among target terms such as female and male, blacks, and whites belonging to various domains like gender and race [36]. Taking it for granted as normal, people rehearse most of these preconceptions in day-to-day discourse, consequently routinizing these linguistic discrimination and making them be felt less visible [76]. Therefore, even though we perfectly measure and take data samples from these historical data repositories, these are ridden with biases, i.e., *Data Bias*, a representative of historical bias, which thereby brings about bias in PLMs [106].

Data bias stemming from innate historical biases is the most general source of bias among different sources of bias explored in literature for various tasks [28], where quality issues in data, uneven distribution (occurrence or co-occurrences) of key terms in data associated with targets concerning a domain [18], etc., are other factors that contribute toward it. Standard datasets used for pre-training non-contextual [16, 22, 40, 67] and contextual models [109] are found to exhibit bias or imbalance in various domains like gender, race, etc. In the context of large PLMs, data bias may be *Pre-trained Data Bias* at the initial training process of PLM or/and *Fine-tuning Data Bias* which is just downstream to it. Studies report that data biases can propagate and get further amplified by underlying machine learning models leading to *Model*

Learning Bias at self-supervised learning strategy to learn linguistic properties and *Downstream Task Learning Bias* at the level of the task-specific fine-tuning model.

Model learning bias is reflected in word representations derived from PLMs and produce *Representation Bias*. Non-contextual word embeddings such as word2vec and GloVe are known to comprise representational biases across gender [16, 22], racial [67] and ethnic groups [40]. It has been clearly unveiled that these embeddings relate word representations of professions like *'nurse'*, *'receptionist'*, and *'homemaker'* to women and *'doctor'*, *'philosopher'*, and *'computer programmer'* to men [16]. Further, they place the representation of words like *'wisdom'* close to representation of *'grandfather'* than *'grandmother'* and encode association of popular African American names with unpleasant phrases [22], etc. Examination of bias in contextualized embeddings reveal that they also exhibit bias like conventional embeddings [18, 125]. For example, BERT is found to encode human-like biases [55]. Most of the recent NLP information retrieval systems such as search engines and question answering highly rely on these biased representations consequently leading to highly biased retrieval behavior. Similarly, use of the large PLM GPT-3 in language generation shows religion bias analogizing *'Muslims'* to *'Terrorists'* [2]. All such biases are disturbing when one observes that word embeddings, being the base constituent of most language systems, can propagate or even intensify them [22, 60, 126] causing unfavorable outcomes when deployed in a plethora of downstream applications such as sentiment analysis [78], language generation [65], and toxic language detection [48].

Downstream task learning bias can cause their outcomes delivered to the public to finally end up in socioeconomic exclusions reinforcing harmful societal stereotypes [16, 60, 126]. Since downstream applications are generally implemented by initializing learning models with an existing source network representation pre-trained on large datasets and later fine-tuned using datasets that suit downstream target task, pre-training data bias, pre-trained representation bias, and fine-tuning data bias all can be the sources to induce bias in downstream applications. Sentiment analysis [52], abusive language detection [80], text classification [33], machine translation [38, 111], personalized medicine [89], coreference resolution [94, 127], crime recidivism prediction systems [27], automating resume screening [56], online advertisements delivery [56, 108], etc., are some of the downstream applications that reports bias in various domains. Inappropriate or unfair choice of label usages to fine-tune downstream task is another source of bias, i.e., *Measurement Bias* [106].

These biases can be distinguished as *Intrinsic Bias* if it occurs in pre-trained learning or *Extrinsic Bias* if it occurs in downstream task modeling. Besides above-mentioned biases, in the perspective of real-world machine learning models, the final system must consider *Evaluation Bias* that occurs when benchmark dataset for a task does not represent certain groups (e.g., images of non-white women not being identified by the face recognition model [20]) and *Deployment Bias* that occurs due to incompatibility of a model designed for a particular task when used differently (e.g., using risk assessment tool created to predict future crime for the purpose of determining length of sentence) [106]. Table 2 shows a large set of research that identify and mitigate bias from large PLMs, where most of them (around 53%) focus primarily on bias identification.

3 Quantifying Bias in PLMs

In order to study unfavorable consequences of different sources of biases in various domains, it is necessary to quantify bias in some manner. Based on the stages of occurrence, bias is generally quantified in the corpus, representation, and downstream tasks.

3.1 Quantifying Bias in Corpora

Counting occurrences or co-occurrences of key terms and deriving various statistics from them help to discover bias in corpora. Bordia and Bowman [18] study gender bias in three publicly available datasets, viz. Penn Treebank [68], WikiText-2 [72], and CNN/Daily Mail [46] that are used to build language models by finding bias scores built using word-level probability profiles within the context of gendered words by defining fixed and infinite-sized context windows around the gendered words. Their bias score helps to find whether words more frequently co-occur with female- or male-gendered words. For fixed sized context windows, they find optimal sized windows as smaller windows can focus more on target words whereas larger windows can focus on the broader topic. Infinite-sized windows are much more stable by using exponentially diminishing weights as the distance between key terms and gendered words increases. Tan et al. [109] count occurrences of key terms (e.g., female or male pronouns) and their co-occurrence with stereotypically gendered occupation terms and perform statistical analysis to find gender bias and also racial and intersectional biases in 1 Billion Word Benchmark [26], BookCorpus [130], Wikipedia, and WebText [88] datasets used to pre-train contextual word models.

3.2 Quantifying Bias in Representations

Geometry of vector spaces Certain works quantify bias by analyzing subspaces in embeddings. Bolukbasi et al. [16] demonstrates the occurrence of gender bias in word embeddings by viewing the difference between word vectors of gendered words like 'sister', 'brother', 'grandmother', 'grandfather', etc., and gender-neutral words. They observe a direction that essentially captures gender and projecting gender neutral words in this direction help them to quantify gender bias. Manzini et al. [67] extends this approach to suit multiclass settings and demonstrate that word representations exhibit several stereotypical biases including religion-based and racial bias.

Table 2 Works addressing bias in large pre-trained language models

Work	PLM	Domain	Quantification	Mitigation
[101]	BERT, XLNet, ALBERT, DistilBERT, RoBERTa, GPT-2	Gender	WEAT, sequence likelihood, pronoun ranking	WEAT scores as an additional loss regularizer
[3]	BERT	Ethnicity	Normalized probability, categorical bias score	M-BERT, contextual word alignment
[112]	DistilBERT, RoBERTa, XLM-RoBERTa, ALBERT, BERT	Gender	Skew and stereotype matrices	Data augmentation
[39]	BERT	Profession, gender	Model querying to predict pronouns at masked position, given context	Gender equality prompt
[13]	BERT	Gender	MLP regressor	Removing word vector components from gender directions
[64]	GPT-2	Politics	Indirect and direct bias metrics	Reinforcement learning
[48]	RoBERTa	Ethnicity, gender	Classification performance, group identifier bias metrics, African American Vernacular English, dialect bias metrics, gender stereotype metrics	Upstream bias mitigation
[97]	GPT-2, T5	Occupation, nationality, religion, gender, age, race, disability	Self-diagnosing	Self-debiasing
[2]	GPT-3	Religion	Prompt completion, analogical reasoning, story generation	–
[103]	XLM, XLM-RoBERTa, BERT multilingual	Gender	Point-wise mutual information and its extension with latent sentiment and regularization	–
[51]	BERT, RoBERTa, ALBERT	Gender, race, sexual orientation, religion, nationality, disability, age, physical appearance, socio-economic status	All unmasked likelihood (AUL), AUL with attention weights	–
[11]	EMLo	Gender	Gender direction in gender subspace	–
[122]	BERT	Gender	Prediction probability scores	–
[43]	ELMo, GPT, BERT, GPT-2	Intersectional	Intersectional bias detection, emergent intersectional bias detection, CEAT	–
[66]	DistilBERT, GPT-2, GPT-NEO	Intersectional bias	Difference/similarity between sentiment scores	–

(continued)

Table 2 (continued)

Work	PLM	Domain	Quantification	Mitigation
[118]	BERT, GPT-2, T5, XLNet	Low frequency names	SV-WEAT, intra and inter layer self-similarity score	–
[61]	BERT	Gender	BERT attention maps	–
[53]	GPT-2	Intersectional	Predictions of GPT-2	–
[75]	BERT, RoBERTa, XLNET, GPT2	Profession, gender, religion, race	Language modelling score, stereotype score, iCAT	–
[65]	LSTM	Gender	Bias score, intervention matches	Counterfactual data augmentation, word embedding debiasing
[124]	BERT pre-trained on medical notes	Ethnicity, gender, insurance groups	Log probability score, extrinsic evaluation using downstream tasks	–
[9]	BERT	Profession, gender	Association test	Name based counterfactual data substitution
[115]	GPT-2	Gender	Causal mediation analysis	–
[109]	GPT-2, CBoW-GLoVe, ELMo, GPT, BERT	Intersectional, gender, race	Counting occurrences, modified SEAT	–
[55]	BERT	Gender	Log probability bias score, WEAT	–
[125]	ELMo	Gender	Co-occurrence	Train-time data augmentation, test-time neutralization
[85]	LSTM	Gender	Occurrences, causal testing, Euclidean distance	Loss function modification
[18]	LSTM	Gender	Fixed and infinite context bias scores	Loss function modification
[10]	ELMo	Gender	Direct bias, biased words clustering and classification	–
[47]	Transformer-XL	Country, occupation, name	Individual and group fairness metrics from sentiment scores	Embedding and sentiment regularization

Word association test Word association tests to quantify representation bias are inspired from the Implicit Association Tests (IAT) in psychology [42] that tries to understand human subconscious bias by measuring differences in the association of target concepts with an attribute. Caliskan et al. [22] proposes a statistical test called the Word Embedding Association Test (WEAT) analogous to IAT to quantify bias in non-contextual word embeddings GloVe and Word2Vec. Using WEAT, authors examine the similarity of embeddings of words in complementary categories like European American and African American names with the complementary attributes like pleasant and unpleasant attributes. The dissimilarity between the association of

European American names with the attributes when compared to African American names with the same attributes helps their study to report the existence of human-like implicit bias in embeddings. To test bias in sentence encoders like ELMo and BERT, May et al. [69] proposes a generalization of WEAT named the Sentence Encoder Association Test (SEAT). Using SEAT, even though they could verify presence of bias in these embeddings, the results were not very generalizable. They also point out that dissimilarities in the results do not mean contextual embeddings is free of bias, rather it may be an indication of cosine similarity not suitable to measure the similarity of embeddings of recent contextual models, and hence, an alternative might be required to quantify bias in such representations. Kurita et al. [55] also shows that conventional cosine similarity-based methods do not produce consistent results to find bias in sentence embeddings generated from the contextual models, as the embeddings of the words may differ according to the context and state of the language model. As the former tests concentrate only on individual words and predefined stereotypical attributes like pleasant and unpleasant terms in artificial contexts that do not reflect the natural use of words, Nadeem et al. [75] proposes two Context Association Tests (CAT) to intrinsically estimate bias in a set of large PLMs. They perform CAT at the sentence level and discourse level, where each target term has natural context, and observe that the recent contextual models BERT, GPT-2, RoBERTa, and XLNet exhibit strong biases with respect to their language modeling ability. Building upon WEAT, Guo, and Caliskan [43] propose Contextualized Embedding Association Test (CEAT) to confirm and extensively quantify biases in neural language models ELMo, BERT, GPT, and GPT-2 according to different contexts.

3.3   Quantifying Bias in Downstream Tasks

In this category, bias is quantified by checking performance scores of system over evaluation corpus that differs only in the context of target terms in which domain of bias is being studied, for example, gender swapping to change gender of gendered words, like '*She is here*' to '*He is here*', and then evaluating model performance of these two sentences [52, 65, 94, 127]. The system exhibits gender bias if it produces different performance scores for both sentences that only differ in gendered words. Dixon et al. [33] presents two performance evaluation metrics derived from error rate equality difference to quantify bias in text classifier constructed to identify toxic comments. Park et al. [80] utilizes these metrics to quantify bias along with the method to generate gender-unbiased dataset proposed by Dixon et al. [33] to find gender bias in abusive language detection. Kiritchenko and Mohammad [52] utilize difference in predicted intensity scores of sentences that differ in gendered words and race, over their corpus to statistically evaluate gender and racial bias in 219 sentiment analysis models that took part in *SemEval 2018: Task 1 Affect In Tweets*. Zhao et al. [127] illustrates gender bias in coreference systems using F1 score over their evaluation corpus named WinoBias that contains pair of sentences that associate gendered pronouns (her/him) to various female or male stereotypical occupations (secretary/physician). Rudinger et al. [94] also perform a similar gender bias study in coreference system over pairs of sentences that differ only in gendered words. Lu et al. [65] quantifies gender bias in coreference resolution and language modeling by calculating dissimilarity in performances of gendered words with various occupations across pairs of sentences that are gender swapped. In addition, bias can also be measured using interpretability, by investigating model interpretations on how it reaches to certain decisions or predictions [35].

4 Mitigating Bias in PLMs

Various efforts have been made for mitigating bias in PLMs to reduce or remove their discriminatory influences over underrepresented or non-mainstream groups. We classify debiasing approaches into three categories, based on the stages at which they are addressed.

4.1 Mitigating Bias in Training Corpora

Most computational algorithms designed for extrinsic applications like sentiment analysis, text classification, etc., primarily rely on labeled training corpora making them prone to societal biases present in these corpora [110]. Despite this being the case, such corpora are still being extensively utilized in various applications of NLP as it is expensive or labor-intensive to build new large-sized training corpora. Hence, generally, common techniques such as data augmentation and bias fine-tuning are followed for debiasing training corpora.

Data augmentation Data augmentation techniques debias the training corpus by supplying additional data to support the target groups with comparatively fewer data in the corpora and thereby creating a balanced corpus to train on. Augmentation can thus in a way counterbalance the under/over-representations of any particular target group of a domain on training corpus. Zhao et al. [125, 127] adopts data augmentation to reduce gender bias in the downstream task of coreference resolution while observing that the possibility of associating occupations to masculine terms is very much higher than feminine terms. They augment gender-swapped versions of data, from masculine terms to feminine terms and vice versa, to the training corpus after anonymizing the named entities. Park et al. [80] utilizes the idea of data augmentation proposed by Zhao et al. [127] to reduce gender bias in abusive language detection. Very similar to Zhao et al. [127], Lu et al. [65] proposes Counterfactual Data Augmentation to explore gender bias in Neural NLP applications including coreference resolution and language modeling by representing bias with reference to internal scores in neural models. Apart from the attempts to remove gender bias in English language, Zmigrod et al. [131] proposes variation to naive gender swapping-based counterfactual data augmentation to manage gender bias in morphology-rich or inflected languages like Spanish and Hebrew that otherwise deliver ungrammatical sentences with simpler approaches. Maudslay et al. [45] improves counterfactual data augmentation by proposing two variants named Counterfactual Data Substitution and Names Intervention to address indirect bias. Liu et al. [63] use concept of Maudslay et al. [45] and propose Counterpart Data Augmentation to remove biases in dialoge systems. Even though data augmentation-based debiasing is a simple technique to reduce bias in the training corpora, their annotation is expensive and increases the size of dataset, in turn, increasing the time required for training. Moreover, these techniques generally only consider isolated words to perform binary swapping and mostly ignore non-binary and more sophisticated representations in a domain. Also, these techniques rely on a predefined limited list of key terms and associated pairs, which may be conceivably incomplete, where some terms may have different spelling (e.g., *mommy* vs. *mummy*), different morphology (e.g., *his*→ *her* and *his*→ *hers*), pairing variations (e.g., *breastfeed*), or produce absurd sentences due to blind swaps (e.g., *she is pregnant*→ *he is pregnant*).

Bias fine-tuning Another approach to debias training corpora is bias fine-tuning by transfer learning, where, instead of the expensive process of constructing balanced corpora with respect to a domain, the idea of transfer learning is employed. As data biases generally originate from data imbalance or small-sized datasets, to make sure the model does not overfit on biased data, Park et al. [80] utilize transfer learning to gender debias abusive language detection model by regularizing from a source network trained on a less biased large dataset and thereafter fine-tune on a target dataset that is largely gender biased. Even though the approach significantly reduces bias, it seems to hurt the overall model accuracy too (as may be expected). Saunders et al. [96] proposes an approach almost opposite to Park et al. [80] where they execute small-domain adaptation by fine-tuning on a less biased small dataset to address gender bias in neural machine translations.

4.2 Mitigating Bias in Representation

An earlier approach to debias the representations is geometric debiasing that removes subspace of protected domain/target terms concerned to a domain in embedding. For a domain, this post-process approach identifies subspace or direction in embedding that holds bias and removes the association of neutral words to target words concerning that domain. This hard debiasing approach is followed by Bolukbasi et al. [16] to alleviate gender bias from word embeddings by leveling the distance of gender-neutral words toward the set of gendered words. They also propose soft bias correction that aims to conserve the initial embedding distances using a trade-off parameter to balance with debiasing. But as it works by removing gender information from the words, it might not be a generalizable approach such as in social science and medical applications that make use of this gender information [8, 70]. Also, their approach makes use of classifier to distinguish gender-neutral words that in turn can propagate classification errors, influencing the performance of debiasing [128]. Zhao et al. [128] neutralize embedding with respect to gender by locating gender-neutral words along with the process of training word vectors without employing an additional classifier. However, Gonen and Goldberg's [41] experiments show that the approaches of Bolukbasi et al. [16] and Zhao et al. [128] are insufficient blind debiasing techniques that just hide but do not actually remove bias. For word-level language models, Bordia and Bowman [18] make use of a loss function regularizer to penalize the projection of embeddings onto gender subspace as a soft debiasing version of Bolukbasi et al. [16]. As stated by Gonen and Goldberg [41], they also mention chances of bias even after debiasing the representations, as a case that their bias score may be not able to detect it. Alternative to this direct geometric debiasing, Zhao et al. [125] propose a different gender neutralization procedure by averaging the representations of original and corresponding gender-swapped data versions for debiasing contextualized representations. To mitigate sentiment bias in word embedding concerned to demographic domains, Sweeney et al. [107] removes the correlations of demographic terms with sentiments.

4.3 Mitigating Bias in Algorithm

Debiasing representations are harder to be applied in contextual embeddings, since the representation of a word can vary according to the plurality of different contexts it appears in [66]. In such cases, certain works focus at modifying algorithm/model, i.e., pre-trained or fine-tuned models, in such a way that it mitigates the bias in predictions during training (i.e., in-processing), usually by modifying the loss function of the model. Qian et al. [85] alters the loss function of the language model in such a way that the model adapts to equalize prediction probabilities for gendered pairs of words, thereby reducing output gender bias. Silva et al. [101] consider utilizing WEAT for debasing transformer-based PLMs and later use WEAT score along with cross entropy to modify loss function of RoBERTa. Huang et al. [47] propose another variation of altering regular cross entropy loss function with embedding and sentiment-driven regularization terms for mitigating sentiment bias in large PLMs. Adversarial training is another way to debias algorithms by altering loss functions while training a predictor along with an adversary, which is intended to produce fair prediction by minimizing the adversarial function that tries to model protected domains like gender, race, etc. [123]. A similar approach is utilized by Zhang et al. [124] to mitigate gender bias in clinical contextual word embeddings. A different recent approach proposed by Liu et al. [64] uses reinforcement learning to debias political bias in large PLMs by utilizing rewards from word embedding or a classifier.

5   Affective Bias

Textual affective computing involves development of algorithms that accurately identify what is subjectively written in text and how to make better affect-influenced subjective decisions, i.e., decisions that are based on emotions, sentiments, or opinions, in several real-world systems. Many systems were initially proposed to detect and measure affect (emotion or sentiment) expressed by textual data such as in movie reviews and product reviews [79, 104]. Later, these systems were widely adopted into various domains such as health care [44], commercial applications [54, 91, 104], politics [21], education [34, 104], and many more. When industrial giants like Google[3], IBM[4] and Microsoft[5] developed natural language understanding tools, textual affective understanding became a crucial and significant part of them. Several observations and arguments have been made by researchers to demonstrate the importance of textual affective computing tasks, especially toward sentiment analysis. According to Cambria et al. [23], sentiment analysis task still needs to travel much to reach human-level performance which can be attained by successfully solving several other NLP problems such as POS tagging and Named Entity Recognition. Poria et al. [83] draw attention to optimistic future research directions in affecting computing like multi-modal affective computing, sarcasm analysis, and even the contemporary issue of social bias in affective NLP systems, and strive to oppose the conventional belief that sentiment analysis tasks are saturated being 20 years old. Our study on affective bias in NLP is motivated from these observations on heightened relevance of affective computing and its wide applicability in diversified NLP applications that leverage affects information [4, 5, 37].

5.1   Definition and Implications

The study on bias in NLP and machine learning was heavily accelerated through observations on the negative impact of NLP biases toward certain social or marginalized groups when applied in the real world[6]. Affective bias is another recent research category in this direction, where researchers study any unfair or imbalanced associations of affect with key terms representing particular underrepresented or protected groups in a domain and how it influences NLP systems, such as sentiment and emotion detection. For example, Google sentiment analyzer is observed to infer that *being gay*[7] is bad and assigns high negative sentiments toward sentences such as *'I'm a gay black woman'* and *'I'm a homosexual'*. By the term Affective bias in NLP, we

---

[3] https://cloud.google.com/natural-language/docs/analyzing-sentiment.

[4] https://cloud.ibm.com/apidocs/natural-language-understanding#emotion.

[5] https://docs.microsoft.com/en-us/azure/cognitive-services/language-service/sentiment-opinion-mining/overview.

[6] https://www.propublica.org/article/Smachine-bias-risk-assessments-in-criminal-sentencing.

[7] https://www.vice.com/en/article/j5jmj8/google-artificial-intelligence-bias.

define the existence of unfair or biased associations of affect (maybe emotions such as anger, fear, and joy or sentiment such as positive, neutral, and negative) toward underrepresented groups, or over-generalized beliefs (stereotypes) about particular social groups in textual documents. For example, in textual documents, words associated with women such as '*she*' and '*wife*' are highly associated with a certain category of emotions like '*sadness*' and '*anger*', and representations of the Muslim religion are observed as being associated with negative terms that indicate violence [1], etc. The existence of affective bias in textual affective computing systems harms its utility and applicability in tasks like business and commercial decision-making, health care, etc. The concept of affective bias is valid and applicable beyond the NLP frameworks because, as in [20], there are chances of high classification error rates for facial emotion detection systems toward underrepresented social groups. Similarly, it is crucial to evaluate all kinds of affective computing systems in the backdrop of affective bias, since automated emotion detection systems have a huge impact on modeling human behavior in many intelligent artificial artifacts or algorithms that imitate human emotion systems for their completeness.

5.2 Affective Bias in PLMs

Many textual affective computing systems including lexicon-based [93], conventional machine learning [107], deep learning [14, 99], and hybrid [32] approaches perpetuate affective bias, which, in general, is transmitted from emotional bias of humans through models learned over large-scale textual corpora. For example, usage of a corpus that contains textual data where humans had expressed the *anger* emotion toward a particular underrepresented religion for training an algorithm can later propagate or/and amplify this affect-oriented bias stereotyping religion with emotion *anger*. Table 3 illustrates an extensive snapshot of works in this area of research along with their major characteristics. A predominant part of existing works study affective bias specific to gender and through the perspective of sentiment analysis [14, 52, 93, 99, 107]. Whereas, other domains like religion, politics, intersectional biases, etc., and their impact in perspective of fine-grained emotion classes (anger, fear, joy sadness, surprise, disgust) have not been investigated as much, except in [52, 114]. The inadequacy of generic evaluation corpora to perform affective bias evaluation in various domains along with the evaluation sentences having corresponding emotion/sentiment ground truth or output label is a notable gap in this area of research. When most works in literature try to identify the existence of affective bias in NLP systems, only very few explore mitigation of the harms of affective bias. The body of work in affective bias can be split into two categories, conventional approaches and runtime verification approaches. Conventional way of analyzing affective bias tries to identify bias in training data, algorithm, representation, and fine-tuning data and, finally, mitigates them using different strategies [52, 107, 114]. Whereas, runtime verification approaches monitor and uncover biased predictions in each run of a specified system using mutations of input sentences (automatically generated templates

Table 3 Works addressing affective bias

Work	Domain	Quantification	Mitigation	Model
Sentiment perspective				
[121]	Gender	BiasFinder in [6]	–	BERT
[6]	Occupation, country of origin, gender	Metamorphic testing	–	BERT, RoBERTa, ALBERT, ELECTRA, Muppet
[107]	Gender	Directional sentiment vectors	Adversarial learning	SVM, LSTM
[93]	Ethnicity, age, sociodemographic status, physical appearance, religion, politics, gender	Projecting word embeddings to cultural axis	–	Lexicon based
[14]	Occupation, gender	Statistical significance difference	–	BERT, Bi-LSTM, logistic regression
[32]	Age	Multinomial log-liner regression, paired t-test	–	Lexicon based, conventional machine learning, hybrid
[99]	Gender, race, politics	Difference in mean sentiment score, statistical significance test, linear regression	–	Rule based, Naive Bayes, dynamic CNN, LSTM
Emotion perspective				
[114]	Gender, race	Mean score of prediction, linear regression on sentiment scores	–	DistilBERT, TextBlob, Google API, VADER
[129]	Non-native English speaker	Wilcoxon signed rank test	Lexical score	VADER, Afinn, SentimentR, TextBlob
[52]	Gender, race	Average score difference	–	Deep learning, conventional machine learning, lexicon based

from input sentences and their paired sentences that represent different views of a stereotype). Runtime verification is generally suitable to validate whether the system satisfies fairness criteria over each run.

The conventional approach by Shen et al. [99] investigates bias in sentiment prediction for textual write-ups comprising similar content generated by different groups of people. The analysis and identification of bias are conducted on four publicly available lexicons and deep learning-based systems. A similar approach by Zhiltsova et al. [129] also identifies and mitigates sentiment bias against non-native English text by using four popular lexicon-based emotion prediction systems. Both these works rely on linguistic style changes across different human groups and how it leads to affective bias in NLP. Apart from analysis of affective bias in lexicon and conventional machine learning systems, researchers also explore non-contextual word embeddings such as word2vec, GloVe, and FastText in the context of affective bias [32, 93, 107]. A significant contribution in this regard is the work by Diaz et al. [32] addressing age-related affective bias in ten widely used word embeddings and fifteen different sentiment analysis models. The work primarily validates whether opinion polling systems falsely report any age group (old or young) more negatively or positively, for example, a sentence with adjectives of '*young*' more likely scores positive sentiments than the same sentence with adjectives of '*old*' [32]. Among the similar studies based on non-contextualized word embeddings, Sweeney et al. [107] introduce an adversarial learning strategy to mitigate demographic affective bias in word2vec and GloVe, and Rozado et al. [93] screen word embeddings to identify bias through the notion of representing words along cultural axis in the embedding space. A more generic work of Kiritchenko et al. [52] identify affective bias in two hundred emotion prediction systems that participated in the shared task *SemEval-2018 Task 1 Affect in Tweets*. They procure an evaluation corpus, Equity Evaluation Corpus (EEC), one of the few publicly available evaluation corpus that has generic evaluation sentences and ground-truth emotion labels as basic emotions anger, fear, joy, and sadness for all evaluation sentences in corpus. A similar evaluation corpus has been proposed by Venkit et al. [114] considering sentences from the domain of persons with disabilities.

A more noteworthy approach to quantify affective bias concerning occupational stereotypes in contextualized large PLM, BERT, is discussed in [14]. Even though more works recently identify and mitigate generic bias in large PLMs due to their efficacy and utility [62, 75], very few of them investigate affective bias in large PLMs. Notable works to uncover bias in sentiment analysis systems that utilize popular PLMs such as Google BERT, Facebook RoBERTa, Google ALBERT, Google ELECTRA, and Facebook Muppet rely on runtime verification approach instead of conventional paradigms to analyze bias [6, 121]. Another interesting approach by Huang et al. [47] investigates sentiment bias introduced in text generated by language models. This emerging scenario facilitates to conduct more evaluations to identify and mitigate affective bias in large PLMs such as BERT, ALBERT, RoBERTa, XLNet, and GPT.

6 Bias Evaluation Corpora

Ideal bias evaluation corpora are indispensable components that help to identify and measure the existence of different types of bias in NLP systems. Since this review focuses on bias in PLMs with special emphasis on affective bias, we present popular and publicly available bias evaluation corpora along with their limitations in the backdrop of PLMs and affective biases, in Table 4; some of the evaluation corpora are avoided due to their unavailability in the public domain to aid future research (e.g., [47, 129]). The most common attempts to create bias evaluation corpora are by initially building template sentences and interchanging its key terms (e.g., *she→ he*, *wife→ husband, mother→ father*) associated with a target (e.g., Female, Male) concerning a domain (e.g., Gender) which induces bias in natural language. For example, Diaz et al. [32] creates bias evaluation corpora intended to identify age-related sentiment bias using template sentences like *'This ⟨ AGE RELATED KEY TERM ⟩ guy was 3 or 4 ft from the tide line and the tide was going out'* and interchanging ⟨ AGE RELATED KEY TERM ⟩ with *'old'* and *'young'*. Contrary to conventional *interchanging of key terms* approach, StereoSet [75] is a much diverse, large-scale, and natural bias evaluation corpus procured through inter- and intra-sentence Context Association Test using Amazon Mechanical Turk. Design and creation of such bias evaluation corpora are often defined carefully to isolate and extract the real effect of bias in various stages such as data bias and algorithm bias, and also to avoid presence of any biased data within evaluation corpus or unbalanced set of key terms referring to a particular target in a domain.

Evaluating large PLMs has recently become highly essential particularly due to their increasing usage in many real-world applications. Several existing bias evaluation corpora are useful enough to evaluate large PLMs when considering the total number, i.e., size, of available evaluation sentences in these corpora [33, 52, 127]. But these corpora fail to represent diverse real-world contexts as they are generally built synthetically using very simple and short-length sentences that are far from real-world scenarios and may even deliver nonsensical samples (e.g., *'my ⟨ sister ⟩ is pregnant'→ 'my ⟨ brother ⟩ is pregnant'*). To accurately design an efficient bias evaluation corpora that suit evaluating PLMs, Liang et al. [62] leverage naturally occurring text corpora from WIKITEXT-2 [72], SST [102], etc., to generate evaluation sentences that can represent real-world context better than the corpus in [33, 52, 75, 100, 127]. There is also a need for more efficient evaluation corpora to evaluate affective bias; one of the available corpus by Kiritchenko et al. [52] contains, as mentioned earlier, very simple synthetic and far from real-world templates where emotions are specified in the sentence explicitly, e.g., *'My brother made me feel angry'* with ground-truth label *anger*. Another affect-oriented corpus by Venkit et al. [114] much suitably represents real-world context suitable to evaluate PLMs; however, they are specifically designed to address bias concerned to people with disabilities, e.g., *'They were aggravated because of the Mentally Handicapped neighbour'* with ground-truth label *anger*. To effectively test and build fair machine learning software, it is highly essential to develop more generic and real-world context-enabled

Table 4 Bias evaluation corpora

Corpora	Domain	Model	Limitation
BITS [114] https://github.com/PranavNV/BITS	Gender, race	Emotion detection	In context of people with disabilities, not generic
[62] https://github.com/pliang279/LM_bias	Gender, religion	Large PLM	–
StereoSet [75] https://github.com/moinnadeem/StereoSet	Gender, race, religion, profession	Large PLM	Less diverse to represent real-world contexts
BEC-Pro [9] https://github.com/marionbartl/gender-bias-BERT	Gender, profession	Large PLM	Less diverse to represent real-world contexts
GenderCorpus [14] https://github.com/jayadevbhaskaran/gendered-sentiment	Gender, profession	Sentiment analysis	Groud-truth sentiment labels are not available
AgeBias [32] https://dataverse.harvard.edu/dataset.xhtml?persistentId=doi:10.7910/DVN/F6EMTS	Age	Sentiment analysis	Groud-truth sentiment labels are not available
EEC [52] https://saifmohammad.com/WebPages/Biases-SA.html	Race, gender	Emotion detection	Explicit representation of emotions, small and simple sentences, less diverse to represent real-world contexts
Winograd [94] https://github.com/rudinger/winogender-schemas	Gender, profession	Coreference resolution	Small sentences
WinoBias [127] https://github.com/uclanlp/corefBias/tree/master/WinoBias/wino	Gender, profession	Coreference resolution	Miss ambiguous pronouns in sufficient volume or diversity to accurately represent practical utility of models
GAP [116] https://github.com/google-research-datasets/gap-coreference	Gender	Coreference resolution	–
Identity synthetic dataset [33] https://github.com/conversationai/unintended-ml-bias-analysis	Human identity (white, gay, etc.)	Text classification	Small and simple sentences, less diverse to represent real-world contexts

evaluation corpora addressing various domains like religion, race, intersectional bias, non-binary representations of gender (including lesbian, gay, etc., apart from common way of only addressing gender as a binary domain). Such corpora can eventually lead to the usage of bias evaluation corpora as an integral part of any NLP system to evaluate system bias like any other test performed on the system before being deployed in real world [24].

7 Discussion on Research Challenges

Here, we discuss several challenges in NLP bias, specifically in the context of large PLMs and also in the context of affective bias.

Heterogeneous nature of NLP bias A large number of research approaches to identify and mitigate NLP biases generally address gender bias (as can be observed from Table 1), even though there exist many other domains such as race, religion, and intersectional biases that need to be addressed significantly. Tan et al. [109] provides evidence in this context that racial bias is encoded strongly in contextualized large PLMs, probably even more than gender bias. They also show that the less explored domain of intersectional biases that consider a mixture of two or more domains (e.g., African American Females) is even more higher than primary biases. Also, most existing research only concentrates on addressing a subset of target terms concerned to a domain, e.g., considering gender domain with binary targets terms male and female, instead of its non-binary nature comprising other target terms like gay and lesbian. All these observations illustrate that the heterogeneous nature of bias in different perspectives must be considered in future works of evaluating bias to make the NLP systems fully debiased.

Evaluation corpus Many existing bias evaluation corpora demonstrate the scarcity of realism in sentences, such as synthetic sentences that are poor in representing real-world context and short-length sentences. Accurate understanding and quantification of existing bias in NLP models may not be effective with these corpora, especially in case of large PLMs, which are capable of generating complex real-world sentences to produce benchmark results in many downstream applications. Most of the large real-world evaluation corpora focus on gender domain, and real-world context-based corpora that suit other domains are scarce. In case of quantification and mitigation of affective biases, many researchers use generic evaluation corpora in [14, 32] where evaluations do not have an affective output label. A corpus developed by Kiritchenko et al. [52] contains relevant emotion labels but lacks real-world contexts and has labels largely derived from explicit mentions of emotions.

Generalizable design Generalizability of any system enables it to be used directly or indirectly within other allied systems. Bias evaluation is becoming an essential part in NLP systems to produce fair decisions by avoiding serious societal harms. This necessity can be achieved largely by designing generalizable bias identification and mitigation modules. Generalizability in the context of evaluation corpora, metrics to identify and quantify the impact of bias, and performance trade-off measures post-debiasing are important challenges to be addressed in the future to bring the vital bias evaluation process as a common and simple component that any NLP system can adapt and experiment.

Pre-train versus Fine-tune bias Bias in downstream applications that utilize pre-trained models originate across different stages of system development. Initially, it may be from massive amount of data used to pre-train the model which may get amplified by the pre-training algorithm. The other way of introducing bias is through fine-tuning data as well as fine-tuning algorithm. An existing challenge in this background is to analyze and understand which part of the overall system really causes bias (i.e., is it from pre-trained or task-specific fine-tuning modules), how bias is perpetuated with the harmonized workflow of system components, which parts of overall system must undergo a mitigation process, etc.

Linguistic diversity and NLP bias Recent NLP research widely explores the utility of developing large PLMs and downstream applications in various different languages by including capability to handle monolingual as well as bilingual strategies. Transformer-based large PLM, IndicBERT[8] which handles low resource and morphologically highly inflected Indian languages like Malayalam, is an example. Similarly, a large variety of monolingual and bilingual corpora[9] (in languages English, Afrikaans, Arabic, etc.) and machine learning models[10] (for various tasks including question answering, text classification, etc.) in different languages illustrate same trends. In this context, it is highly essential to study bias in such different linguistic corpora and the mono and multilingual PLMs learned from those corpora. Among the very few attempts in this direction is the research to mitigate gender bias in Hindi word embedding [84], evaluating gender bias in Hindi–English machine translation [90], addressing gender bias in languages like Spanish and Hebrew [131], etc.

Interdisciplinary methods to analyze NLP bias Sun et al. [105] states that many techniques deployed in non-NLP systems can be applied directly or with small modifications to identify and mitigate bias in NLP. Several studies indicate that bias in NLP systems opens up scope to associate with other branches such as sociology, psychology, socio-linguistics, engineering, and legal studies. The Implicit Association Test [42] in psychology and its diverse computational representations such as Word Embedding Association Test [22] and Sentence Encoder Association Test [69] to quantify representation bias are examples to showcase interdisciplinary nature of removing bias in NLP. Also, lessons from Software Engineering[11] such as unit testing and behavior testing can be adopted to evaluate and quantify bias in machine

---

[8] https://huggingface.co/ai4bharat/indic-bert, accessed on January 03 2022.

[9] https://huggingface.co/datasets, accessed on January 03 2022.

[10] https://huggingface.co/models, accessed on January 03 2022.

[11] "IndoML-2021- Day 1 Session 1", YouTube video, 01:12:50, Posted by "CSE IIT Gandhinagar" December, 16, 2021, https://www.youtube.com/watch?v=y3t8pc1s0Yw, accessed on January 03 2022.

learning models at different levels [92]. Sun et al. [105] treats mitigation of NLP bias as a combined problem of sociology and engineering, where sociology can identify how really humans perceive and encode social bias into language. Research toward interdisciplinary discussions can bring light into current bias quantification and mitigation strategies and inspire developing advanced and practically relevant approaches [7, 12, 98].

8 Conclusion

In this survey, we conducted a comprehensive investigation toward bias in large pre-trained language models, especially the transformer-based models. We discuss different types of biases that originate at various stages of pre-trained language model workflow and the methods used to quantify and mitigate these biases. Due to the widespread utility of affective computing systems in the real world, we give a special emphasis on the less explored area of bias that associates with affect, i.e., affective bias. Our study also lists the popular and publicly available evaluation corpora that aid future research, along with their suitability in large pre-trained language models. Finally, we discuss the challenges in this area of research, addressing which would aid the community to further improve the tasks of identification and mitigation of bias in pre-trained language modes. Materials regarding this survey will be made publicly available at https://github.com/anoopkdcs/NLPBias, and we will keep updating it to aid future research.

Acknowledgements The author 'Manjary P. Gangan' was Supported by the Women Scientist Scheme-A (WOS-A) for Research in Basic/Applied Science from the Department of Science and Technology (DST) of the Government of India under the Grant SR/WOS-A/PM-62/2018.

References

1. Abid A, Farooqi M, Zou J (2021) Large language models associate Muslims with violence. Nat Mach Intell 3(6):461–463. https://doi.org/10.1038/s42256-021-00359-2
2. Abid A, Farooqi M, Zou J (2021) Persistent anti-Muslim bias in large language models. Association for Computing Machinery, New York, NY, USA, pp 298–306. https://doi.org/10.1145/3461702.3462624
3. Ahn J, Oh A (2021) Mitigating language-dependent ethnic bias in BERT. In: Proceedings of the 2021 conference on empirical methods in natural language processing. Association for Computational Linguistics, pp 533–549. https://doi.org/10.18653/v1/2021.emnlp-main.42
4. Anoop K (2019) Affect-oriented fake news detection using machine learning. In: AWSAR awarded popular science stories by scientists for the people. Vigyan Prasar, DST, India, pp 426–428. ISBN: 978-81-7480-337-5. https://www.researchgate.net/publication/344838679_Affect-Oriented_Fake_News_Detection_Using_Machine_Learning
5. Anoop K, Deepak P, Lajish VL (2020) Emotion cognizance improves health fake news identification. In: Proceedings of the 24th symposium on international database engineering &

applications. IDEAS '20. Association for Computing Machinery, Seoul, Republic of Korea. https://doi.org/10.1145/3410566.3410595

6. Asyrofi MH, Yang Z, Yusuf INB, Kang HJ, Thung F, Lo D (2021) Biasfinder: metamorphic test generation to uncover bias for sentiment analysis systems. IEEE Trans Softw Eng. https://doi.org/10.1109/TSE.2021.3136169

7. Avin C, Keller B, Lotker Z, Mathieu C, Peleg D, Pignolet YA (2015) Homophily and the glass ceiling effect in social networks. In: Proceedings of the 2015 conference on innovations in theoretical computer science, ITCS '15. Association for Computing Machinery, New York, NY, USA, pp 41–50. https://doi.org/10.1145/2688073.2688097

8. Back SE, Payne RL, Simpson AN, Brady KT (2010) Gender and prescription opioids: findings from the national survey on drug use and health. Addict Behav 35(11):1001–1007. https://doi.org/10.1016/j.addbeh.2010.06.018

9. Bartl M, Nissim M, Gatt A (2020) Unmasking contextual stereotypes: measuring and mitigating BERT's gender bias. In: Proceedings of the second workshop on gender bias in natural language processing. Association for Computational Linguistics, pp 1–16. https://aclanthology.org/2020.gebnlp-1.1

10. Basta C, Costa-jussà MR, Casas N (2019) Evaluating the underlying gender bias in contextualized word embeddings. In: Proceedings of the first workshop on gender bias in natural language processing. Association for Computational Linguistics, Italy, pp 33–39. https://doi.org/10.18653/v1/W19-3805

11. Basta C, Costa-jussà MR, Casas N (2021) Extensive study on the underlying gender bias in contextualized word embeddings. Neural Comput Appl 33(8):3371–3384. https://doi.org/10.1007/s00521-020-05211-z

12. Beukeboom CJ, Burgers C (2019) How stereotypes are shared through language: a review and introduction of the social categories and stereotypes communication (SCSC) framework. Rev Commun Res 7:1–37. https://doi.org/10.12840/issn.2255-4165.017

13. Bhardwaj R, Majumder N, Poria S (2021) Investigating gender bias in Bert. Cogn Comput 1–11. https://doi.org/10.1007/s12559-021-09881-2

14. Bhaskaran J, Bhallamudi I (2019) Good secretaries, bad truck drivers? Occupational gender stereotypes in sentiment analysis. In: Proceedings of the first workshop on gender bias in natural language processing. Association for Computational Linguistics, Italy, pp 62–68. https://doi.org/10.18653/v1/W19-3809

15. Blodgett SL, Green L, O'Connor B (2016) Demographic dialectal variation in social media: a case study of African-American English. In: Proceedings of the 2016 conference on empirical methods in natural language processing. Association for Computational Linguistics, Austin, TX, pp 1119–1130. https://doi.org/10.18653/v1/D16-1120

16. Bolukbasi T, Chang KW, Zou J, Saligrama V, Kalai A (2016) Man is to computer programmer as woman is to homemaker? Debiasing word embeddings. In: Proceedings of the 30th international conference on neural information processing systems. NIPS'16, Curran Associates Inc., pp 4356–4364

17. Bolukbasi T, Chang KW, Zou J, Saligrama V, Kalai A (2016) Quantifying and reducing stereotypes in word embeddings. arXiv:1606.06121

18. Bordia S, Bowman SR (2019) Identifying and reducing gender bias in word-level language models. In: Proceedings of the 2019 conference of the North American chapter of the Association for Computational Linguistics: student research workshop. Association for Computational Linguistics, Minneapolis, MN, pp 7–15. https://doi.org/10.18653/v1/N19-3002

19. Brown T, Mann B, Ryder N, Subbiah M, Kaplan JD, Dhariwal P, Neelakantan A, Shyam P, Sastry G, Askell A, Agarwal S, Herbert-Voss A, Krueger G, Henighan T, Child R, Ramesh A, Ziegler D, Wu J, Winter C, Hesse C, Chen M, Sigler E, Litwin M, Gray S, Chess B, Clark J, Berner C, McCandlish S, Radford A, Sutskever I, Amodei D (2020) Language models are few-shot learners. In: Advances in neural information processing systems, vol 33. Curran Associates, Inc., pp 1877–1901. https://proceedings.neurips.cc/paper/2020/file/1457c0d6bfcb4967418bfb8ac142f64a-Paper.pdf

20. Buolamwini J, Gebru T (2018) Gender shades: intersectional accuracy disparities in commercial gender classification. In: Proceedings of the 1st conference on fairness, accountability and transparency. Proceedings of machine learning research, vol 81. PMLR, pp 77–91. https://proceedings.mlr.press/v81/buolamwini18a.html

21. Caetano JA, Lima HS, Santos MF, Marques-Neto HT (2018) Using sentiment analysis to define Twitter political users' classes and their homophily during the 2016 American presidential election. J Internet Serv Appl 9(1):1–15. https://doi.org/10.1186/s13174-018-0089-0

22. Caliskan A, Bryson JJ, Narayanan A (2017) Semantics derived automatically from language corpora contain human-like biases. Science 356(6334):183–186. https://doi.org/10.1126/science.aal4230

23. Cambria E, Poria S, Gelbukh A, Thelwall M (2017) Sentiment analysis is a big suitcase. IEEE Intell Syst 32(6):74–80. https://doi.org/10.1109/MIS.2017.4531228

24. Chakraborty J, Majumder S, Yu Z, Menzies T (2020) Fairway: a way to build fair ML software. In: Proceedings of the 28th ACM joint meeting on European software engineering conference and symposium on the foundations of software engineering. Association for Computing Machinery, New York, NY, USA, pp 654–665. https://doi.org/10.1145/3368089.3409697

25. Chaloner K, Maldonado A (2019) Measuring gender bias in word embeddings across domains and discovering new gender bias word categories. In: Proceedings of the first workshop on gender bias in natural language processing. Association for Computational Linguistics, Italy, pp 25–32. https://doi.org/10.18653/v1/W19-3804

26. Chelba C, Mikolov T, Schuster M, Ge Q, Brants T, Koehn P (2013) One billion word benchmark for measuring progress in statistical language modeling. Computing Research Repository (CoRR) pp 1–6

27. Chouldechova A (2017) Fair prediction with disparate impact: a study of bias in recidivism prediction instruments. Big Data 5(2):153–163. https://doi.org/10.1089/big.2016.0047. pMID: 28632438

28. Corbett-Davies S, Pierson E, Feller A, Goel S, Huq A (2017) Algorithmic decision making and the cost of fairness. In: Proceedings of the 23rd ACM SIGKDD international conference on knowledge discovery and data mining, KDD '17. Association for Computing Machinery, New York, NY, USA, pp 797–806. https://doi.org/10.1145/3097983.3098095

29. Craft JT, Wright KE, Weissler RE, Queen RM (2020) Language and discrimination: generating meaning, perceiving identities, and discriminating outcomes. Annu Rev Linguist 6(1):389–407. https://doi.org/10.1146/annurev-linguistics-011718-011659

30. Dev S, Li T, Phillips JM, Srikumar V (2020) On measuring and mitigating biased inferences of word embeddings. In: Proceedings of the AAAI conference on artificial intelligence, vol 34(05), pp 7659–7666. https://ojs.aaai.org/index.php/AAAI/article/view/6267

31. Devlin J, Chang MW, Lee K, Toutanova K (2019) BERT: pre-training of deep bidirectional transformers for language understanding. In: Proceedings of the 2019 conference of the North American chapter of the Association for Computational Linguistics: human language technologies, vol 1 (long and short papers). Association for Computational Linguistics, Minneapolis, MN, pp 4171–4186. https://doi.org/10.18653/v1/N19-1423

32. Díaz M, Johnson I, Lazar A, Piper AM, Gergle D (2018) Addressing age-related bias in sentiment analysis. In: Proceedings of the 2018 chi conference on human factors in computing systems. Association for Computing Machinery, New York, NY, USA, pp 1–14. https://doi.org/10.1145/3173574.3173986

33. Dixon L, Li J, Sorensen J, Thain N, Vasserman L (2018) Measuring and mitigating unintended bias in text classification. In: Proceedings of the 2018 AAAI/ACM conference on AI, ethics, and society, AIES '18. Association for Computing Machinery, New York, NY, USA, pp 67–73. https://doi.org/10.1145/3278721.3278729

34. Dolianiti FS, Iakovakis D, Dias SB, Hadjileontiadou S, Diniz JA, Hadjileontiadis L (2018) Sentiment analysis techniques and applications in education: a survey. In: International conference on technology and innovation in learning, teaching and education. Springer, pp 412–427. https://doi.org/10.1007/978-3-030-20954-4_31

35. Du M, Yang F, Zou N, Hu X (2021) Fairness in deep learning: a computational perspective. IEEE Intell Syst 36(4):25–34. https://doi.org/10.1109/MIS.2020.3000681
36. Eagly A, Wood W, Diekman A (2000) Social role theory of sex differences and similarities: a current appraisal. Lawrence Erlbaum Associates Publishers, pp 123–174
37. Elmadany A, Zhang C, Abdul-Mageed M, Hashemi A (2020) Leveraging affective bidirectional transformers for offensive language detection. In: Proceedings of the 4th workshop on open-source Arabic Corpora and processing tools, with a shared task on offensive language detection. European Language Resource Association, France, pp 102–108. https://aclanthology.org/2020.osact-1.17
38. Escudé Font J, Costa-jussà MR (2019) Equalizing gender bias in neural machine translation with word embeddings techniques. In: Proceedings of the first workshop on gender bias in natural language processing. Association for Computational Linguistics, Italy, pp 147–154. https://doi.org/10.18653/v1/W19-3821
39. Fatemi Z, Xing C, Liu W, Xiong C (2021) Improving gender fairness of pre-trained language models without catastrophic forgetting. arXiv:2110.05367
40. Garg N, Schiebinger L, Jurafsky D, Zou J (2018) Word embeddings quantify 100 years of gender and ethnic stereotypes. Proc Natl Acad Sci 115(16):E3635–E3644
41. Gonen H, Goldberg Y: Lipstick on a pig: debiasing methods cover up systematic gender biases in word embeddings but do not remove them. In: Proceedings of the 2019 conference of the North American chapter of the Association for Computational Linguistics: human language technologies, vol 1 (long and short papers). Association for Computational Linguistics, Minneapolis, MN, pp 609–614 (2019). https://doi.org/10.18653/v1/N19-1061
42. Greenwald AG, McGhee DE, Schwartz JL (1998) Measuring individual differences in implicit cognition: the implicit association test. J Pers Soc Psychol 74(6):1464
43. Guo W, Caliskan A (2021) Detecting emergent intersectional biases: contextualized word embeddings contain a distribution of human-like biases. In: Proceedings of the 2021 AAAI/ACM conference on AI, ethics, and society. Association for Computing Machinery, New York, NY, USA, pp 122–133. https://doi.org/10.1145/3461702.3462536
44. Gupta VS, Kohli S (2016) Twitter sentiment analysis in healthcare using Hadoop and R. In: 2016 3rd international conference on computing for sustainable global development (INDIACom). IEEE, pp 3766–3772. https://ieeexplore.ieee.org/document/7724965
45. Hall Maudslay R, Gonen H, Cotterell R, Teufel S (2019) It's all in the name: mitigating gender bias with name-based counterfactual data substitution. In: Proceedings of the 2019 conference on empirical methods in natural language processing and the 9th international joint conference on natural language processing (EMNLP-IJCNLP). Association for Computational Linguistics, Hong Kong, China, pp 5267–5275. https://doi.org/10.18653/v1/D19-1530
46. Hermann KM, Kočiský T, Grefenstette E, Espeholt L, Kay W, Suleyman M, Blunsom P (2015) Teaching machines to read and comprehend. In: Proceedings of the 28th international conference on neural information processing systems, vol 1, pp 1693–1701
47. Huang PS, Zhang H, Jiang R, Stanforth R, Welbl J, Rae J, Maini V, Yogatama D, Kohli P (2020) Reducing sentiment bias in language models via counterfactual evaluation. In: Findings of the Association for Computational Linguistics: EMNLP 2020. Association for Computational Linguistics, pp 65–83 (online). https://doi.org/10.18653/v1/2020.findings-emnlp.7
48. Jin X, Barbieri F, Kennedy B, Davani AM, Neves L, Ren X (2021) On transferability of bias mitigation effects in language model fine-tuning. In: Proceedings of the 2021 conference of the North American chapter of the Association for Computational Linguistics: human language technologies. Association for Computational Linguistics, pp 3770–3783. https://doi.org/10.18653/v1/2021.naacl-main.296
49. Joulin A, Grave E, Bojanowski P, Douze M, Jégou H, Mikolov T (2016) Fasttext. zip: compressing text classification models. arXiv:1612.03651
50. Kalyan KS, Rajasekharan A, Sangeetha S (2021) Ammus: a survey of transformer-based pretrained models in natural language processing. arXiv:2108.05542

51. Kaneko M, Bollegala D (2021) Unmasking the mask—evaluating social biases in masked language models. arXiv:2104.07496
52. Kiritchenko S, Mohammad S (2018) Examining gender and race bias in two hundred sentiment analysis systems. In: Proceedings of the seventh joint conference on lexical and computational semantics. Association for Computational Linguistics, New Orleans, Louisiana, pp 43–53. https://doi.org/10.18653/v1/S18-2005
53. Kirk HR, Volpin F, Iqbal H, Benussi E, Dreyer F, Shtedritski A, Asano Y et al (2021) Bias out-of-the-box: an empirical analysis of intersectional occupational biases in popular generative language models. Advances in neural information processing systems, vol 34
54. Krishnamoorthy S (2018) Sentiment analysis of financial news articles using performance indicators. Knowl Inf Syst 56(2):373–394. https://doi.org/10.1007/s10115-017-1134-1
55. Kurita K, Vyas N, Pareek A, Black AW, Tsvetkov Y (2019) Measuring bias in contextualized word representations. In: Proceedings of the first workshop on gender bias in natural language processing. Association for Computational Linguistics, Italy, pp 166–172. https://doi.org/10.18653/v1/W19-3823
56. Lambrecht A, Tucker C (2019) Algorithmic bias? An empirical study of apparent gender-based discrimination in the display of stem career ads. Manag Sci 65(7):2966–2981. https://doi.org/10.1287/mnsc.2018.3093
57. Lapowsky I (2018) Google autocomplete still makes vile suggestions
58. Le Q, Mikolov T (2014) Distributed representations of sentences and documents. In: Proceedings of the 31st international conference on international conference on machine learning, ICML'14, vol 32. JMLR.org, pp II-1188–II-1196. https://dl.acm.org/doi/10.5555/3044805.3045025
59. Lee J, Yoon W, Kim S, Kim D, Kim S, So CH, Kang J (2019) BioBERT: a pre-trained biomedical language representation model for biomedical text mining. Bioinformatics 36(4):1234–1240. https://doi.org/10.1093/bioinformatics/btz682
60. Leino K, Fredrikson M, Black E, Sen S, Datta A (2019) Feature-wise bias amplification. In: International conference on learning representations. https://openreview.net/forum?id=S1ecm2C9K7
61. Li B, Peng H, Sainju R, Yang J, Yang L, Liang Y, Jiang W, Wang B, Liu H, Ding C (2021) Detecting gender bias in transformer-based models: a case study on Bert. arXiv:2110.15733
62. Liang PP, Wu C, Morency LP, Salakhutdinov R (2021) Towards understanding and mitigating social biases in language models. In: International conference on machine learning. PMLR, pp 6565–6576. http://proceedings.mlr.press/v139/liang21a.html
63. Liu H, Dacon J, Fan W, Liu H, Liu Z, Tang J (2020) Does gender matter? Towards fairness in dialogue systems. In: Proceedings of the 28th international conference on computational linguistics. International Committee on Computational Linguistics, Barcelona, Spain, pp 4403–4416 (online). https://doi.org/10.18653/v1/2020.coling-main.390
64. Liu R, Jia C, Wei J, Xu G, Wang L, Vosoughi S (2021) Mitigating political bias in language models through reinforced calibration. Proc AAAI Conf Artif Intell 35(17):14857–14866. https://ojs.aaai.org/index.php/AAAI/article/view/17744
65. Lu K, Mardziel P, Wu F, Amancharla P, Datta A (2020) Gender bias in neural natural language processing. Logic, language, and security. Springer, Cham, pp 189–202
66. Magee L, Ghahremanlou L, Soldatic K, Robertson S (2021) Intersectional bias in causal language models. arXiv:2107.07691
67. Manzini T, Yao Chong L, Black AW, Tsvetkov Y (2019) Black is to criminal as Caucasian is to police: detecting and removing multiclass bias in word embeddings. In: Proceedings of the 2019 conference of the North American chapter of the Association for Computational Linguistics: human language technologies, vol 1 (long and short papers). Association for Computational Linguistics, Minneapolis, MN, pp 615–621. https://doi.org/10.18653/v1/N19-1062
68. Marcus MP, Santorini B, Marcinkiewicz MA (1993) Building a large annotated corpus of English: the Penn Treebank. Comput Linguist 19(2):313–330. https://aclanthology.org/J93-2004

69. May C, Wang A, Bordia S, Bowman SR, Rudinger R (2019) On measuring social biases in sentence encoders. In: Proceedings of the 2019 conference of the North American chapter of the Association for Computational Linguistics: human language technologies, vol 1 (long and short papers). Association for Computational Linguistics, Minneapolis, MN, pp 622–628. https://doi.org/10.18653/v1/N19-1063

70. McFadden AC, Marsh GE, Price BJ, Hwang Y (1992) A study of race and gender bias in the punishment of school children. Education and treatment of children, pp 140–146

71. Menegatti M, Rubini M (2017). Gender bias and sexism in language. https://doi.org/10.1093/acrefore/9780190228613.013.470

72. Merity S, Xiong C, Bradbury J, Socher R (2016) Pointer sentinel mixture models. arXiv:1609.07843

73. Mikolov T, Chen K, Corrado G, Dean J (2013) Efficient estimation of word representations in vector space. In: 1st international conference on learning representations, ICLR 2013, Scottsdale, Arizona, USA, 2–4 May 2013, workshop track proceedings

74. Motro D, Evans J, Ellis AP (2019) Benson L (2019) Race and reactions to negative feedback: Examining the effects of the "angry black woman" stereotype. Acad Manag Proc 2019(1):11230. https://doi.org/10.5465/AMBPP.2019.11230abstract

75. Nadeem M, Bethke A, Reddy S (2021) StereoSet: measuring stereotypical bias in pretrained language models. In: Proceedings of the 59th annual meeting of the Association for computational Linguistics and the 11th international joint conference on natural language processing (volume 1: long papers). Association for Computational Linguistics, pp 5356–5371 (online). https://doi.org/10.18653/v1/2021.acl-long.416

76. Ng SH (2007) Language-based discrimination: blatant and subtle forms. J Lang Soc Psychol 26(2):106–122. https://doi.org/10.1177/0261927X07300074

77. Niethammer C (2020) Ai bias could put women's lives at risk—a challenge for regulators. https://www.forbes.com/sites/carmenniethammer/2020/03/02/ai-bias-could-put-womens-lives-at-riska-challenge-for-regulators/?sh=753a6217534f

78. Packer B, Mitchell M, Guajardo-Céspedes M, Halpern Y (2018) Text embeddings contain bias. Here's why that matters. Tech rep, Google

79. Pang B, Lee L, Vaithyanathan S (2002) Thumbs up? Sentiment classification using machine learning techniques. In: Proceedings of the ACL-02 conference on empirical methods in natural language processing, EMNLP '02, vol 10. Association for Computational Linguistics, USA, pp 79–86. https://doi.org/10.3115/1118693.1118704

80. Park JH, Shin J, Fung P (2018) Reducing gender bias in abusive language detection. In: Proceedings of the 2018 conference on empirical methods in natural language processing. Association for Computational Linguistics, Brussels, Belgium, pp 2799–2804. https://doi.org/10.18653/v1/D18-1302

81. Pennington J, Socher R, Manning C (2014) GloVe: global vectors for word representation. In: Proceedings of the 2014 conference on empirical methods in natural language processing (EMNLP). Association for Computational Linguistics, Doha, Qatar, pp 1532–1543. https://doi.org/10.3115/v1/D14-1162

82. Peters ME, Neumann M, Iyyer M, Gardner M, Clark C, Lee K, Zettlemoyer L (2018) Deep contextualized word representations. In: Proceedings of the 2018 conference of the North American chapter of the Association for Computational Linguistics: human language technologies, vol 1 (long papers). Association for Computational Linguistics, New Orleans, Louisiana, pp 2227–2237. https://doi.org/10.18653/v1/N18-1202

83. Poria S, Hazarika D, Majumder N, Mihalcea R (2020) Beneath the tip of the iceberg: current challenges and new directions in sentiment analysis research. IEEE Trans Affect Comput. https://doi.org/10.1109/TAFFC.2020.3038167

84. Pujari AK, Mittal A, Padhi A, Jain A, Jadon M, Kumar V (2019) Debiasing gender biased Hindi words with word-embedding. In: Proceedings of the 2019 2nd international conference on algorithms, computing and artificial intelligence. Association for Computing Machinery, New York, NY, USA, pp 450–456. https://doi.org/10.1145/3377713.3377792

85. Qian Y, Muaz U, Zhang B, Hyun JW (2019) Reducing gender bias in word-level language models with a gender-equalizing loss function. In: Proceedings of the 57th annual meeting of the Association for Computational Linguistics: student research workshop. Association for Computational Linguistics, Italy, pp 223–228. https://doi.org/10.18653/v1/P19-2031
86. Qiu X, Sun T, Xu Y, Shao Y, Dai N, Huang X (2020) Pre-trained models for natural language processing: a survey. Sci China Technol Sci 1–26. https://doi.org/10.1007/s11431-020-1647-3
87. Radford A, Narasimhan K, Salimans T, Sutskever I (2018) Improving language understanding by generative pre-training. OpenAI blog
88. Radford A, Wu J, Child R, Luan D, Amodei D, Sutskever I et al (2019) Language models are unsupervised multitask learners. OpenAI blog 1(8):9
89. Rajkomar A, Hardt M, Howell MD, Corrado G, Chin MH (2018) Ensuring fairness in machine learning to advance health equity. Ann Intern Med 169(12):866–872. https://doi.org/10.7326/M18-1990 pMID: 30508424
90. Ramesh K, Gupta G, Singh S (2021) Evaluating gender bias in Hindi-English machine translation. In: Proceedings of the 3rd workshop on gender bias in natural language processing. Association for Computational Linguistics, pp 16–23. https://doi.org/10.18653/v1/2021.gebnlp-1.3
91. Renault T (2020) Sentiment analysis and machine learning in finance: a comparison of methods and models on one million messages. Digit Finance 2(1):1–13. https://doi.org/10.1007/s42521-019-00014-x
92. Ribeiro MT, Wu T, Guestrin C, Singh S (2020) Beyond accuracy: behavioral testing of NLP models with CheckList. In: Proceedings of the 58th annual meeting of the Association for Computational Linguistics. Association for Computational Linguistics, pp 4902–4912 (online). https://doi.org/10.18653/v1/2020.acl-main.442
93. Rozado D (2020) Wide range screening of algorithmic bias in word embedding models using large sentiment lexicons reveals underreported bias types. PLOS ONE 15(4):1–26. https://doi.org/10.1371/journal.pone.0231189
94. Rudinger R, Naradowsky J, Leonard B, Van Durme B (2018) Gender bias in coreference resolution. In: Proceedings of the 2018 conference of the North American chapter of the Association for Computational Linguistics: human language technologies, vol 2 (short papers). Association for Computational Linguistics, New Orleans, Louisiana, pp 8–14. https://doi.org/10.18653/v1/N18-2002
95. Sambasivan N, Arnesen E, Hutchinson B, Doshi T, Prabhakaran V (2021) Re-imagining algorithmic fairness in India and beyond. In: Proceedings of the 2021 ACM conference on fairness, accountability, and transparency, FAccT '21. Association for Computing Machinery, New York, NY, USA, pp 315–328. https://doi.org/10.1145/3442188.3445896
96. Saunders D, Byrne B (2020) Reducing gender bias in neural machine translation as a domain adaptation problem. In: Proceedings of the 58th annual meeting of the Association for Computational Linguistics. Association for Computational Linguistics, pp 7724–7736 (online). https://doi.org/10.18653/v1/2020.acl-main.690
97. Schick T, Udupa S, Schütze H (2021) Self-diagnosis and self-debiasing: a proposal for reducing corpus-based bias in NLP. Trans Assoc Comput Linguist 9:1408–1424. https://doi.org/10.1162/tacl_a_00434
98. Schluter N (2018) The glass ceiling in NLP. In: Proceedings of the 2018 conference on empirical methods in natural language processing. Association for Computational Linguistics, Brussels, Belgium, pp 2793–2798. https://doi.org/10.18653/v1/D18-1301
99. Shen JH, Fratamico L, Rahwan I, Rush AM (2018) Darling or babygirl? Investigating stylistic bias in sentiment analysis. In: Proceedings of FATML
100. Sheng E, Chang KW, Natarajan P, Peng N (2019) The woman worked as a babysitter: on biases in language generation. In: Proceedings of the 2019 conference on empirical methods in natural language processing and the 9th international joint conference on natural language processing (EMNLP-IJCNLP). Association for Computational Linguistics, Hong Kong, China, pp 3407–3412. https://doi.org/10.18653/v1/D19-1339

101. Silva A, Tambwekar P, Gombolay M (2021) Towards a comprehensive understanding and accurate evaluation of societal biases in pre-trained transformers. In: Proceedings of the 2021 conference of the North American chapter of the Association for Computational Linguistics: human language technologies. Association for Computational Linguistics, pp 2383–2389. https://doi.org/10.18653/v1/2021.naacl-main.189

102. Socher R, Perelygin A, Wu J, Chuang J, Manning CD, Ng AY, Potts C (2013) Recursive deep models for semantic compositionality over a sentiment treebank. In: Proceedings of the 2013 conference on empirical methods in natural language processing. Association for Computational Linguistics, Seattle, Washington, USA, pp 1631–1642. https://aclanthology.org/D13-1170

103. Stańczak K, Choudhury SR, Pimentel T, Cotterell R, Augenstein I (2021) Quantifying gender bias towards politicians in cross-lingual language models. arXiv:2104.07505

104. Suharshala R, Anoop K, Lajish VL (2018) Cross-domain sentiment analysis on social media interactions using senti-lexicon based hybrid features. In: 2018 3rd international conference on inventive computation technologies (ICICT). IEEE, Coimbatore, India, pp 772–777. https://doi.org/10.1109/ICICT43934.2018.9034272

105. Sun T, Gaut A, Tang S, Huang Y, ElSherief M, Zhao J, Mirza D, Belding E, Chang KW, Wang WY (2019) Mitigating gender bias in natural language processing: literature review. In: Proceedings of the 57th annual meeting of the Association for Computational Linguistics. Association for Computational Linguistics, Italy, pp 1630–1640. https://doi.org/10.18653/v1/P19-1159

106. Suresh H, Guttag J (2021) A framework for understanding sources of harm throughout the machine learning life cycle. In: Equity and access in algorithms, mechanisms, and optimization, EAAMO '21. Association for Computing Machinery, New York, NY, USA. https://doi.org/10.1145/3465416.3483305

107. Sweeney C, Najafian M (2020) Reducing sentiment polarity for demographic attributes in word embeddings using adversarial learning. In: Proceedings of the 2020 conference on fairness, accountability, and transparency, FAT* '20. Association for Computing Machinery, New York, NY, USA, pp 359–368. https://doi.org/10.1145/3351095.3372837

108. Sweeney L (2013) Discrimination in online ad delivery: Google ads, black names and white names, racial discrimination, and click advertising. Queue 11(3):10–29. https://doi.org/10.1145/2460276.2460278

109. Tan YC, Celis LE (2019) Assessing social and intersectional biases in contextualized word representations. In: Advances in neural information processing systems, vol 32. Curran Associates, Inc. https://proceedings.neurips.cc/paper/2019/file/201d546992726352471cfea6b0df0a48-Paper.pdf

110. Torralba A, Efros AA (2011) Unbiased look at dataset bias. In: CVPR 2011. IEEE, pp 1521–1528. https://doi.org/10.1109/CVPR.2011.5995347

111. Vanmassenhove E, Hardmeier C, Way A (2018) Getting gender right in neural machine translation. In: Proceedings of the 2018 conference on empirical methods in natural language processing. Association for Computational Linguistics, Belgium, pp 3003–3008. https://doi.org/10.18653/v1/D18-1334

112. de Vassimon Manela D, Errington D, Fisher T, van Breugel B, Minervini P (2021) Stereotype and skew: quantifying gender bias in pre-trained and fine-tuned language models. In: Proceedings of the 16th conference of the European chapter of the Association for Computational Linguistics: main volume. Association for Computational Linguistics, pp 2232–2242. https://doi.org/10.18653/v1/2021.eacl-main.190

113. Vaswani A, Shazeer N, Parmar N, Uszkoreit J, Jones L, Gomez AN, Kaiser Ł, Polosukhin I (2017) Attention is all you need. In: Advances in neural information processing systems, vol 30. Curran Associates, Inc., pp 5998–6008. https://proceedings.neurips.cc/paper/2017/file/3f5ee243547dee91fbd053c1c4a845aa-Paper.pdf

114. Venkit PN, Wilson S (2021) Identification of bias against people with disabilities in sentiment analysis and toxicity detection models. arXiv:2111.13259

115. Vig J, Gehrmann S, Belinkov Y, Qian S, Nevo D, Singer Y, Shieber S (2020) Investigating gender bias in language models using causal mediation analysis. In: Advances in neural information processing systems, vol 33. Curran Associates, Inc., pp 12388–12401. https://proceedings.neurips.cc/paper/2020/file/92650b2e92217715fe312e6fa7b90d82-Paper.pdf

116. Webster K, Recasens M, Axelrod V, Baldridge J (2018) Mind the GAP: a balanced corpus of gendered ambiguous pronouns. Trans Assoc Comput Linguist 6:605–617. https://doi.org/10.1162/tacl_a_00240

117. Weidinger L, Mellor J, Rauh M, Griffin C, Uesato J, Huang PS, Cheng M, Glaese M, Balle B, Kasirzadeh A et al (2021) Ethical and social risks of harm from language models. arXiv:2112.04359

118. Wolfe R, Caliskan A (2021) Low frequency names exhibit bias and overfitting in contextualizing language models. In: Proceedings of the 2021 conference on empirical methods in natural language processing. Association for Computational Linguistics, and Punta Cana, Dominican Republic, pp 518–532 (online). https://doi.org/10.18653/v1/2021.emnlp-main.41

119. Yang Y, Uy MCS, Huang A (2020) Finbert: a pretrained language model for financial communications. arXiv:2006.08097

120. Yang Z, Dai Z, Yang Y, Carbonell J, Salakhutdinov RR, Le QV (2019) Xlnet: generalized autoregressive pretraining for language understanding. Advances in neural information processing systems, vol 32. https://proceedings.neurips.cc/paper/2019/file/dc6a7e655d7e5840e66733e9ee67cc69-Paper.pdf

121. Yang Z, Asyrofi MH, Lo D (2021) Biasrv: uncovering biased sentiment predictions at runtime. In: Proceedings of the 29th ACM joint meeting on European software engineering conference and symposium on the foundations of software engineering, ESEC/FSE 2021. Association for Computing Machinery, New York, NY, USA, pp 1540–1544. https://doi.org/10.1145/3468264.3473117

122. Ye W, Xu F, Huang Y, Huang C et al (2021) Adversarial examples generation for reducing implicit gender bias in pre-trained models. arXiv:2110.01094

123. Zhang BH, Lemoine B, Mitchell M (2018) Mitigating unwanted biases with adversarial learning. In: Proceedings of the 2018 AAAI/ACM conference on AI, ethics, and society, AIES '18. Association for Computing Machinery, New York, NY, USA, pp 335–340. https://doi.org/10.1145/3278721.3278779

124. Zhang H, Lu AX, Abdalla M, McDermott M, Ghassemi M (2020) Hurtful words: quantifying biases in clinical contextual word embeddings. In: Proceedings of the ACM conference on health, inference, and learning, CHIL '20. Association for Computing Machinery, New York, NY, USA, pp 110–120. https://doi.org/10.1145/3368555.3384448

125. Zhao J, Wang T, Yatskar M, Cotterell R, Ordonez V, Chang KW (2019) Gender bias in contextualized word embeddings. In: Proceedings of the 2019 conference of the North American chapter of the Association for Computational Linguistics: human language technologies, vol 1 (long and short papers). Association for Computational Linguistics, Minneapolis, Minnesota, pp 629–634. https://doi.org/10.18653/v1/N19-1064

126. Zhao J, Wang T, Yatskar M, Ordonez V, Chang KW (2017) Men also like shopping: reducing gender bias amplification using corpus-level constraints. In: Proceedings of the 2017 conference on empirical methods in natural language processing. Association for Computational Linguistics, Copenhagen, Denmark, pp 2979–2989. https://doi.org/10.18653/v1/D17-1323

127. Zhao J, Wang T, Yatskar M, Ordonez V, Chang KW (2018) Gender bias in coreference resolution: evaluation and debiasing methods. In: Proceedings of the 2018 conference of the North American chapter of the Association for Computational Linguistics: human language technologies, vol 2 (short papers). Association for Computational Linguistics, New Orleans, Louisiana, pp 15–20. https://doi.org/10.18653/v1/N18-2003

128. Zhao J, Zhou Y, Li Z, Wang W, Chang KW (2018) Learning gender-neutral word embeddings. In: Proceedings of the 2018 conference on empirical methods in natural language processing. Association for Computational Linguistics, Brussels, Belgium, pp 4847–4853. https://doi.org/10.18653/v1/D18-1521

129. Zhiltsova A, Caton S, Mulway C (2019) Mitigation of unintended biases against non-native English texts in sentiment analysis. In: Proceedings for the 27th AIAI Irish conference on artificial intelligence and cognitive science, Galway, Ireland, 5–6 Dec 2019. CEUR workshop proceedings, vol 2563, pp 317–328. CEUR-WS.org. http://ceur-ws.org/Vol-2563/aics_30.pdf

130. Zhu Y, Kiros R, Zemel R, Salakhutdinov R, Urtasun R, Torralba A, Fidler S (2015) Aligning books and movies: towards story-like visual explanations by watching movies and reading books. In: Proceedings of the 2015 IEEE international conference on computer vision (ICCV), pp 19–27

131. Zmigrod R, Mielke SJ, Wallach H, Cotterell R (2019) Counterfactual data augmentation for mitigating gender stereotypes in languages with rich morphology. In: Proceedings of the 57th annual meeting of the Association for Computational Linguistics. Association for Computational Linguistics, Italy, pp 1651–1661. https://doi.org/10.18653/v1/P19-1161

Exploring Rawlsian Fairness
for K-Means Clustering

Stanley Simoes, Deepak P., and Muiris MacCarthaigh

1 Introduction

While traditional machine learning (ML) algorithms aim to maximise utility through discrimination, they are now being increasingly regulated by law so as to prevent any unjustifiable discrimination in the society, especially when these algorithms are deployed on a large scale and can heavily influence people's life chances. These laws provide the interpretation of fairness (i.e. the *four-fifths rule*[1] in the US and the *Equality Act 2010*[2] in the UK) which existing algorithms are required to comply with. The challenge lies in translating these legal interpretations of fairness, which are in natural language, to mathematical formulations that can be incorporated in ML algorithms with minimal impact on the utility of these algorithms.

Among the many ideas of justice and fairness put forth by political philosophers and social scientists over time, the ML research community has looked at the idea of fairness that is computationally easy to model such as individual fairness [6] and group fairness [14]. However, this idea does not have enough support from the justice and fairness space. We take a novel and different route in this paper: we look at the ideas of *John Rawls*[3]—an influential 20th century moral and political philosopher in liberal tradition—and attempt to incorporate his ideas of fairness in unsupervised ML. Compared to other ideas, the Rawlsian ideas are time-tested, and a good mix

---

[1] https://www.law.cornell.edu/cfr/text/29/1607.4.

[2] https://www.legislation.gov.uk/ukpga/2010/15/contents.

[3] https://plato.stanford.edu/entries/rawls/.

S. Simoes · Deepak P. (✉) · M. MacCarthaigh
Queen's University Belfast, Belfast, UK
e-mail: deepaksp@acm.org

S. Simoes
e-mail: ssimoes01@qub.ac.uk

M. MacCarthaigh
e-mail: M.MacCarthaigh@qub.ac.uk

© The Author(s), under exclusive license to Springer Nature Singapore Pte Ltd. 2022
J. Mathew et al. (eds.), *Responsible Data Science*, Lecture Notes
in Electrical Engineering 940, https://doi.org/10.1007/978-981-19-4453-6_3

of pragmatism and principledness, but incorporating them into ML algorithms is challenging.

In general, ideas of fairness can be incorporated in existing ML algorithms in 3 ways—(i) *preprocessing*: where the input is processed to ensure fairness before being fed to the ML algorithm, (ii) *inprocessing*: where the ML algorithm is altered to incorporate the fairness criteria (e.g. by adding fairness constraints to the optimisation criterion), and (iii) *postprocessing*: where the output of the ML algorithm is processed to make it fairer. Preprocessing and postprocessing techniques can be used with any off-the-shelf ML algorithm with no modification, which is not the case with the inprocessing techniques. Existing ML algorithms can thus be easily augmented to produce fairer outputs through preprocessing and postprocessing techniques. On the other hand, the inprocessing techniques are known to have both better utility and better fairness than the other two.

In this paper, we focus on *clustering*—an unsupervised ML task. We specifically look at the *k-means clustering* algorithm [8], a well-known unsupervised ML algorithm that is used to partition a collection of examples into disjoint sets called *clusters* such that similar examples are assigned to the same cluster. We look at the k-means clustering algorithm with an additional constraint of satisfying Rawls' *difference principle* [15]. In other words, we allow the overall utility of the obtained clusters to be sub-optimal as long as the least-advantaged sensitive group has the greatest utility. We work towards developing a *postprocessing* technique that perturbs the output of the standard k-means clustering algorithm to incorporate the difference principle. While there has been research on the intersection of fairness and clustering, to the best of our knowledge, this is the first to attempt to incorporate Rawlsian ideas in clustering. Although this work is exploratory, we hope that this report of our experiences would aid future work in this direction.

2   The Difference Principle and Rawlsian Point

In his book *Justice as Fairness: A Restatement*, John Rawls states the *difference principle* as

> Social and economic inequalities are to satisfy two conditions: first, they are to be attached to offices and positions open to all under conditions of fair equality of opportunity; and second, they are to be to the greatest benefit of the least-advantaged members of society. [15]

We focus on the second condition: *they [social and economic inequalities] are to be to the greatest benefit of the least-advantaged members of society*. For the experiments outlined in this paper, we interpret the above statement as the members of a society belonging to one of two sensitive social groups (e.g. male or female),[4] one being the more advantaged group and the other the less advantaged group.

---

[4] We use binary genders for sake of illustration only.

2.1 Terminology

In the context of clustering, we refer to the *Rawlsian point* as the cluster assignment where the difference principle is satisfied, i.e. the utility to the least-advantaged members of society is the greatest, and we refer to such a cluster assignment as the *Rawlsian k-means clusters*. In contrast, the classical k-means clustering algorithm by design returns a cluster assignment where the sum of individual utilities is (approximately) maximised; we refer to this point as the *utilitarian point*, and the corresponding cluster assignment as the *utilitarian k-means clusters*. This paper looks at binary sensitive attributes, i.e. we assume that all examples in the dataset belong to one of two sensitive groups. We refer to the sensitive group with the lower utility as the less advantaged group (LAG) and the one with the higher utility as the more advantaged group (MAG). Thus, the LAG and MAG are defined on the sensitive attribute. Note that the sensitive attribute is *not* used in the k-means clustering algorithm.

3 Problem Statement

We attempt to develop a postprocessing technique that operates on the output of the k-means clustering algorithm, returning a cluster assignment that corresponds to the Rawlsian point, i.e. the Rawlsian k-means clusters. This requires that the Rawlsian point does indeed exist. This leads us to the following questions:

1. Does there exist a cluster assignment that corresponds to (or is an approximate of) the Rawlsian point? (Sect. 5)
2. Can we reach this point starting from the utilitarian point, i.e. the point of highest utility achieved by the classical k-means clustering algorithm? (Sect. 6).

3.1 Notation

Let $\mathcal{X} = [\ldots, x, \ldots]$ be a collection of examples defined over a set of non-sensitive attributes $\mathcal{N}$ and a single binary sensitive attribute $S = \{0, 1\}$. For any example x, let $x \cdot n$ denotes its values for the non-sensitive attributes and $x \cdot s$ denotes its value for the sensitive attribute. Also, let $x \cdot n \in [0, 1]^{|\mathcal{N}|}$ and $x \cdot s \in \{0, 1\}$. Note that the non-sensitive attributes $\mathcal{N}$ are the only attributes used for clustering; the clustering algorithm does not use the sensitive attribute S.

The k-means clustering algorithm assigns a label to each example based on the example's distance from the cluster centroids, thus yielding a cluster assignment $\mathcal{C} = \{\ldots, C, \ldots\}$ with $|\mathcal{C}| = k$, where each cluster C is a set of examples having the same label and k is the number of clusters to be generated. We additionally define the *utility* $u(x)$ of an example x as

$$u(x) = \delta - d(x \cdot n, C) \tag{1}$$

where the constant δ is the maximum possible distance between two examples, and $d(x.n, C)$ is the distance of example x from the nearest cluster centroid. Further, the utility of a sensitive group α is computed as

$$U(\alpha) = \frac{1}{|\mathcal{X}_\alpha|} \sum_{x \in \mathcal{X}_\alpha} u(x) \tag{2}$$

where $\mathcal{X}_\alpha$ denotes the set of examples in $\mathcal{X}$ belonging to the sensitive group α, i.e. $\mathcal{X}_\alpha = \{x | x \in \mathcal{X}, x \cdot s = \alpha\}$. The overall utility of a cluster assignment is the average utility of all examples in the dataset.

In the next section, we describe the Adult dataset which is central to this exploratory study, and in the subsequent sections, we attempt to address the questions outlined above.

4 Dataset

We use the publicly available Adult dataset[5] from the UCI repository [5] in our experiments. The Adult dataset has been heavily used in the fairness literature. It consists of 15 attributes and 30718 examples with no missing values. We use 8 non-sensitive attributes, 1 sensitive attribute, and the predictor. Table 1 lists out the attributes from this dataset that are used for preprocessing, clustering, and evaluation in our experiments.

Preprocessing We follow the steps in previous work [1] for preprocessing the dataset. The non-sensitive attributes (used for clustering) are preprocessed as follows:

- continuous: scaled and translated to the range [0, 1].
- categorical: one hot encoded. To ensure that the maximum squared distance between any two examples for this attribute is 1, the value for the 'hot' position is set to $\frac{1}{\sqrt{2}}$.

Consequently, the maximum distance between any two values of a single non-sensitive attribute is 1. As we use 8 non-sensitive attributes, the maximum possible distance δ between any two examples is 8. We then undersample for parity across the predictor attribute. The resulting dataset contains 42 non-sensitive attributes and 1 sensitive attribute. Finally, we sample 500 examples from each predictor class for a total of 1000 examples. Table 2 shows the distribution of sensitive groups in the sampled dataset.

---

[5] https://archive.ics.uci.edu/ml/datasets/adult.

Table 1 Attributes from the Adult dataset used in our experiments

	Type	Name
Sensitive	Categorical	Sex
Non-sensitive	Continuous	Age
		Education num
		Capital gain
		Capital loss
		Hours per week
	Categorical	Workclass
		Education
		Occupation
Predictor	Categorical	Annual income

Only the non-sensitive attributes are used in the clustering algorithm. The sensitive attribute is used only for computing the utilities of the sensitive groups. The predictor attribute is used only when preprocessing the dataset

Table 2 Distribution of sensitive groups (female, male) in the sampled dataset (1000 examples)

Sex	# of examples	% of examples (%)
Female	267	26.7
Male	733	73.3

5 Existence of Rawlsian k-Means Clusters

Before devising a technique for obtaining the Rawlsian k-means clusters, we need to determine whether such a cluster assignment indeed exists. Since it is impractical to enumerate and evaluate the utilities of all possible cluster assignments, we instead perform several runs of the k-means algorithm on our dataset with different initial centroids to find an approximate. Note that either of the sensitive groups may be less (or more) advantaged; in case of the Adult dataset, we only consider those cluster assignments where the minority group (i.e. female) is the less advantaged group. Figure 1 shows the utilities of these cluster assignments, with the utility of the more advantaged group (MAG) on the x-axis and the less advantaged group (LAG) on the y-axis.

It can be seen in Fig. 1 that there is indeed a point (shown as an olive plus point) corresponding to a k-means clustering that has a better utility for the less advantaged group than the utilitarian point (shown as a blue plus point). The olive plus point is thus an approximate for the Rawlsian point. Note that this point may not be the actual Rawlsian point; we use it as an approximate for the Rawlsian point in the experiment discussed in Sect. 6 as evaluating all possible cluster assignments is not possible.

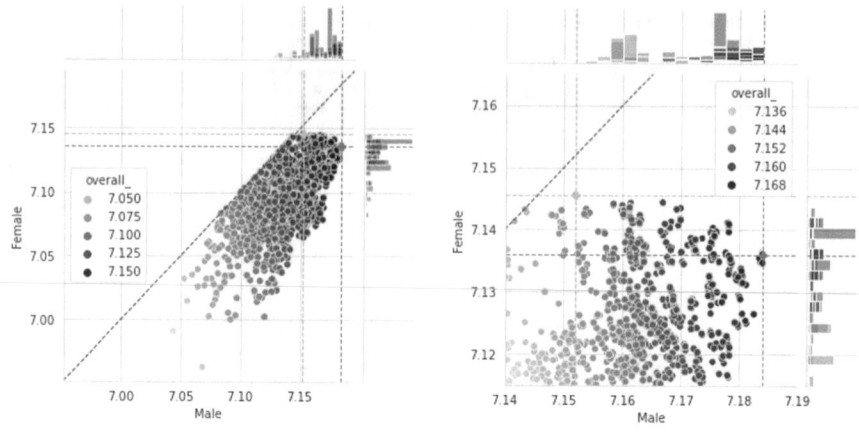

(a) points generated by the 5000 runs

(b) zooming in to the utilitarian point and (approximate) Rawlsian point

Fig. 1 Scatter plot of points in the MAG-LAG utility space generated by 5000 runs of the k-means clustering algorithm with different initial centroids, $k = 5$, on our Adult dataset, and majority (male) as the more advantaged group (MAG) and minority (female) as the less advantaged group (LAG). Each point in this space corresponds to the cluster assignment obtained from a single run. The hue of a point indicates the overall utility, with darker being better. Points on the 45° red dashed line are such that the utilities for MAG and LAG are equal. By design of this experiment, i.e. the minority group (female) being the less advantaged group, all points are always south of this 45° line. The horizontal/vertical olive dashed lines intersect at the point with the highest LAG (female) utility, but has low overall utility since female is a minority group (see Table 2). This olive point is an approximate for the Rawlsian point. The horizontal/vertical blue dashed lines intersect at the point with the highest MAG (male) utility and is also of the highest overall utility (i.e. the utilitarian point) since male is a majority group. The histograms to the top and right of each scatter plot indicate the density of the points

6 Reaching the Approximate Rawlsian Point

We saw in the previous section there is no mechanism to generate points in the MAG-LAG utility space other than through the k-means algorithm itself. We thus performed several runs of the k-means algorithm to see whether doing so can reveal spaces, where the Rawlsian point is likely to be and whether we can navigate to these spaces. Now, the histograms to the top and right of the scatter plots in Fig. 1 indicate that the generated points are concentrated near the utilitarian point (the blue plus point). This would suggest that a k-means algorithm would likely generate a cluster assignment that corresponds to a point closer to the utilitarian point. Thus, we select the utilitarian cluster assignment, i.e. the one with highest overall utility (shown as a blue plus point in Fig. 1) as the starting point for our postprocessing technique.

We now outline our postprocessing technique for arriving at the approximate Rawlsian point starting from the utilitarian point. Our goal is to find a *reassignment* of examples to new clusters so that we reach the approximate Rawlsian point (shown as a olive plus point). To do so, we apply a series of reassignment operations to the utilitarian cluster assignment such that we gradually move above the horizontal blue line in the MAG-LAG utility space towards the Rawlsian point. The general outline of our technique is shown in Algorithm 1. We discuss the GenerateOperations algorithm in Sect. 7, and how the best operation is selected and applied in Sect. 6.1.

Algorithm 1 Traverse

Require: $\mathcal{X}, \mathcal{C}$
Ensure: $\mathcal{C}$
1: **while** $\mathcal{C}$ has not changed **do**
2:      $O \leftarrow$ GenerateOperations$(\mathcal{X}, \mathcal{C})$ ▷ Generate operations (Sect. 7)
3:      $o \leftarrow$ select the best operation in O ▷ (Sect. 6.1)
4:      **if** o is not ϕ **then**
5:          $\mathcal{C} \leftarrow$ apply the selected operation o
6:      **end if**
7: **end while**

6.1 Selecting and Applying the Best Reassignment Operation

We select the best reassignment operation among those generated according to the following order of preference:

1. Among the operations that generate a point in the northeast of the current point in the MAG-LAG utility space, select the operation that corresponds to the point with highest overall utility.
2. If no such operation exists, among the operations that generate a point in the *skyline* of points in the northwest of the current point, select the operation that corresponds to the point with highest overall utility.
3. If no such operation exists, select the null operation ϕ, which indicates that no reassignment is done and hence the cluster assignment is unchanged.

The *skyline* S of a set of points Q is defined as those points in the MAG-LAG utility space where $\forall q \in Q, \exists s \in S$ such that either (i) $q == s$ (i.e. q is in the skyline), or (ii) $u_{LAG}(q) < u_{LAG}(s)$ and $u_{MAG}(q) < u_{MAG}(s)$ (i.e. q is worse than s for both LAG and MAG), where $u_\alpha(x)$ is the utility of sensitive group α for the point x. Moving through the skyline ensures that we select the operation with the least drop in overall utility and maximum gain in LAG utility.

Applying the selected reassignment operation is straightforward; we change the current cluster assignment (i.e. current labels of the examples) as specified by the selected operation, and use the new cluster assignment as the starting point for the next iteration.

7 Reassignment Operators

Our goal is to construct an operator that reassigns a number of examples in the current cluster assignment to new clusters, thus generating a new cluster assignment—the Rawlsian cluster assignment—having a higher LAG utility while minimally affecting the overall utility. By applying a series of instantiations of this operator to the utilitarian cluster assignment, we hope to reach the approximate Rawlsian point. We explore two simple operators $\mathbf{R_1}$ and $\mathbf{R_2}$ which are now detailed.

7.1 Reassignment Operator $\mathbf{R_1}$

The reassignment operator $\mathbf{R_1}$ operating on a tuple (x, C') takes a single example x from the current cluster assignment C and reassigns it to a different cluster C' (i.e. a different label) thus yielding a new cluster assignment. The number of possible operations[6] generated for $\mathbf{R_1}$ is thus $n \times (k - 1)$. Algorithm 2 outlines the GenerateOperationsR1 algorithm which is the $\mathbf{R_1}$ variant of the GenerateOperations algorithm.

Algorithm 2 GenerateOperationsR1

Require: $\mathcal{X}, C$
Ensure: O
1: Initialise O to an empty set
2: **for** each example-cluster tuple $(x, C') \in \mathcal{X} \times C$ **do**
3:    Apply $\mathbf{R_1}(x, C')$ to get a new cluster assignment C'
4:    Obtain the corresponding point in the MAG-LAG utility space
5:    **if** the point has a higher LAG utility than the current point **then**
6:       Add (x, C') to O
7:    **end if**
8:    Discard C'
9: **end for**

Figure 2 shows the trajectory of points in the MAG-LAG utility space generated by employing GenerateOperationsR1 in Algorithm 1. We see that the algorithm takes small steps (each step corresponds to the application of one $\mathbf{R_1}$ operation) in the correct direction towards the approximate Rawlsian point but ends up with the LAG and MAG utilities being equal. Notably, the approximate Rawlsian point (olive plus point in the figure) has a better LAG (female) utility than all generated points. There is thus a need for improvement. Another drawback of $\mathbf{R_1}$ is that its repeated application may be inefficient because the amount of movement towards the approximate Rawlsian point in each iteration is negligible. Additionally, the reassignment operation may never be undone.

---

[6] An *operation* is an instantiated operator.

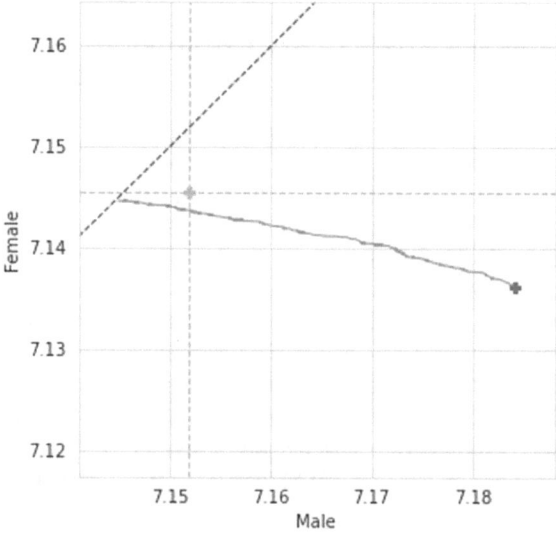

Fig. 2 Trajectory of points (in orange) generated in the MAG-LAG utility space by Traverse (Algorithm 1) using the R_1 operator (Sect. 7.1). Points on the 45° red dashed line are such that the utilities for MAG and LAG are equal. The olive plus point corresponds to the approximate Rawlsian point which we would like the Traverse algorithm to arrive at. Recall that the Rawlsian point corresponds to the cluster assignment, where the utility of the less advantaged group is the greatest. The algorithm starts from the utilitarian point (indicated by the blue plus point) and repeatedly applies a R_1 operation thus yielding a new cluster assignment, with each cluster assignment generating a corresponding point in the utility space. This figure contains 261 such generated points. The orange line that connects the generated sequence of points together indicates the trajectory of the algorithm which is generally moving in the northwest direction, and can be seen to terminate at the red line, i.e. where the utilities for MAG and LAG are equal

7.2 Pair Reassignment Operator R_2

To overcome the inefficiency limitation of R_1, we instead select pairs of examples to be reassigned. We define the *pair reassignment* operator as follows: The pair reassignment operator R_2 operating on a pair of tuples (x_1, C_1'), (x_2, C_2') takes a pair of distinct examples x_1 and x_2 ($x_1 \neq x_2$) from the current cluster assignment C and reassigns them to new clusters C_1' and C_2' thus yielding a new cluster assignment. One R_2 operation is thus equivalent to two R_1 operations on two distinct examples x_1 and x_2.

Two issues arise when using R_2 for navigating the MAG-LAG utility space:

1. The number of possible operations using R_2 is $\binom{n}{2}(k-1)^2$, where n is the number of examples and k is the number of clusters. In our experiment, this evaluates to nearly 8 million possible operations; it is impractical to calculate the utilities of all such operations. In contrast, the number of possible operations using R_1 is $n \times (k-1)$ which is significantly smaller.

2. On inspection of a sample of these R_2 operations, we found that a large chunk (around 62%) generate points with lower LAG utilities than the current point, i.e. they move away from the Rawlsian point, and hence are not useful.

Thus, there is a need to intelligently generate R_2 operations whose corresponding points have higher LAG utilities than the current point. Algorithm 3 outlines the GenerateOperationsR2 algorithm which is the R_2 variant of the GenerateOperations algorithm. Instead of generating all possible operations, we use a heuristic for pruning the set of example-cluster tuples (Lines 1 to 4 in Algorithm 3). This heuristic is based on the assumption that if R_1 does not generate a point with higher LAG utility for some tuple (x, C'), then no R_2 operation instantiated with (x, C') — i.e. neither $R_2(x, C', \cdot, \cdot)$ nor $R_2(\cdot, \cdot, x, C')$ — will generate a point with higher LAG utility than the current point. Next, the remaining tuples are separately ranked on LAG utility and overall utility, and only the top reassignments (1% to 5%) from the two rankings are retained. This yields a set of good example-cluster tuples $T_{good} \subset \mathcal{X} \times \mathcal{C}$ that can be later used to instantiate R_2. The rest of the algorithm is similar to Algorithm 2.

Algorithm 3 GenerateOperationsR2

Require: $\mathcal{X}, \mathcal{C}$
Ensure: O
1: $T \leftarrow$ GenerateOperationsR1$(\mathcal{X}, \mathcal{C})$
2: $T_{LAG} \leftarrow$ top $p\%$ of T ranked on LAG utility
3: $T_{overall} \leftarrow$ top $q\%$ of T ranked on overall utility
4: $T_{good} \leftarrow T_{LAG} \cup T_{overall}$
5: Initialise O to an empty set
6: **for** each pair of example-cluster tuples $((x_1, C_1'), (x_2, C_2')) \in T_{good} \times T_{good}$ **do**
7:      Apply $R_2(x_1, C_1', x_2, C_2')$ to get a new cluster assignment C'
8:      Obtain the corresponding point in the MAG-LAG utility space
9:      **if** the point has a higher LAG utility than the current point **then**
10:         Add (x_1, C_1', x_2, C_2') to O
11:     **end if**
12:     Discard C'
13: **end for**

Figure 3 compares the trajectories of points in the MAG-LAG utility space generated by employing GenerateOperationsR1 and GenerateOperationsR2 in Algorithm 1. We can see that while R_2 improves upon the inefficiency limitation of R_1 (requiring 152 iterations as compared to 261 iterations by R_1), it follows nearly an identical trajectory as R_1 and hence still suffers from its other limitations, i.e. ends up with the LAG and MAG utilities being equal, and the operation once applied may never be undone.

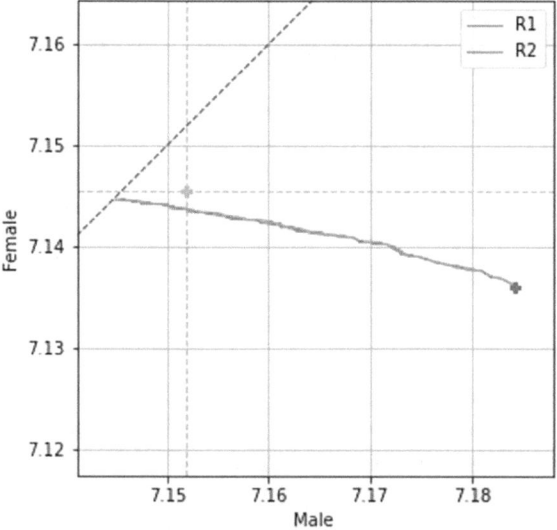

Fig. 3 Trajectories of points generated in the MAG-LAG utility space by Traverse (Algorithm 1) using the R_1 operator (in orange) and the R_2 operator (in green). Points on the 45° red dashed line are such that the utilities for MAG and LAG are equal. The olive plus point corresponds to the approximate Rawlsian point which we would like the Traverse algorithm to arrive at. Recall that the Rawlsian point corresponds to the cluster assignment where the utility of the less advantaged group is the greatest. The algorithm starts from the utilitarian point (indicated by the blue plus point) and repeatedly applies a R_1 (or R_2) operation thus yielding a new cluster assignment, with each cluster assignment generating a corresponding point in the utility space. R_1's trajectory consists of 261 such generated points, and R_2's trajectory consists of 152 such generated points. Similar to R_1's trajectory, R_2 moves in the northwest direction terminating at the red line, i.e. where the utilities for MAG and LAG are equal

8 Related Work

While there has been significant research on fairness in machine learning in recent years, we look at those works that are relevant to this paper, i.e. (i) Rawlsian ideas of fairness in machine learning and fair algorithms for clustering.

Rawlsian ideas of fairness in ML: The few existing works that explore Rawls' ideas of fairness in ML are in the supervised setting. The hardness of adapting Rawlsian principles into algorithms is apparent from these works. For example, Shah et al. [16] propose a classifier that minimises the error rate of the worst-off sensitive group; they call this a Rawls classifier. Hashimoto et al. [7] employ Rawlsian ideas to mitigate the amplification of representation disparity in empirical risk minimization. We have not come across any work that explores Rawls' ideas in the unsupervised setting.

Fair algorithms for clustering: These can be broadly categorised based on the notion of fairness, i.e. (i) individual fairness [9, 11, 12] and (ii) group fairness [3, 4, 10]. Further, these algorithms differ on how the fairness criterion is enforced (i.e.

preprocessing, inprocessing or postprocessing). For example, Chierichetti et al. [3] propose a preprocessing technique that makes the output of a subsequent standard clustering algorithm fair. Kleindessner et al. [10] incorporate the fairness constraint within the clustering algorithm. Davidson and Ravi [4] look at postprocessing clusters for fairness; they do so by presenting it as a minimum cluster modification for group fairness (MCMF) optimisation problem which is formulated as an ILP. Other notions of fairness such as representativity fairness [13] and proportionality fairness [2] have also been proposed for clustering.

9 Conclusion

We proposed a postprocessing framework for making the clusters generated by the standard k-means clustering algorithm satisfy Rawls' difference principle while minimally affecting the overall utility. Within this framework, we explored two simple operators that perturb a given cluster assignment by reassigning examples to new clusters; the first operator R_1 reassigns a single example to a new cluster at a time, and the second operator R_2 reassigns two examples to new clusters at a time. We observed that while R_2 improves upon the efficiency limitation of R_1, there is still a huge scope for improvement with regards to arriving at the Rawlsian point. There is a need to design an operator that reassigns a larger number of examples to other clusters, which will consequently open up new avenues for exploration in the search space. Nevertheless, we expect these operators to act as good baselines for any future operator.

Acknowledgements This project has received funding from the European Union's Horizon 2020 research and innovation programme under the Marie Skłodowska-Curie grant agreement No. 945231; and the Department of the Economy in Northern Ireland.

References

1. Abraham SS, Deepak P, Sundaram SS (2020) Fairness in clustering with multiple sensitive attributes. In: Proceedings of the 23rd international conference on extending database technology. OpenProceedings.org, pp 287–298. https://openproceedings.org/2020/conf/edbt/paper_148.pdf
2. Chen X, Fain B, Lyu L, Munagala K (2019) Proportionally fair clustering. In: Proceedings of the 36th international conference on machine learning. PMLR, pp 1032–1041. https://proceedings.mlr.press/v97/chen19d.html
3. Chierichetti F, Kumar R, Lattanzi S, Vassilvitskii S (2017) Fair clustering through fairlets. In: Advances in neural information processing systems, vol 30. Curran Associates, Inc. https://proceedings.neurips.cc/paper/2017/hash/978fce5bcc4eccc88ad48ce3914124a2-Abstract.html

4. Davidson I, Ravi SS (2020) Making existing clusterings fairer: algorithms, complexity results and insights. In: Proceedings of the thirty-fourth AAAI conference on artificial intelligence, pp 3733–3740. https://ojs.aaai.org//index.php/AAAI/article/view/5783
5. Dua D, Graff C (2017) UCI machine learning repository. http://archive.ics.uci.edu/ml
6. Dwork C, Hardt M, Pitassi T, Reingold O, Zemel R (2012) Fairness through awareness. In: Proceedings of the 3rd innovations in theoretical computer science conference. Association for Computing Machinery, pp 214–226. https://doi.org/10.1145/2090236.2090255
7. Hashimoto T, Srivastava M, Namkoong H, Liang P (2018) Fairness without demographics in repeated loss minimization. In: Proceedings of the 35th international conference on machine learning. PMLR, pp 1929–1938. https://proceedings.mlr.press/v80/hashimoto18a.html
8. Hastie T, Tibshirani R, Friedman J (2009) The elements of statistical learning: data mining, inference, and prediction. Unsupervised learning, 2nd edn. Springer, New York, NY, pp 485–585. https://doi.org/10.1007/978-0-387-84858-7_14
9. Jung C, Kannan S, Lutz N (2019) A center in your neighborhood: fairness in facility location. CoRR. http://arxiv.org/abs/1908.09041
10. Kleindessner M, Awasthi P, Morgenstern J (2019) Fair k-center clustering for data summarization. In: Proceedings of the 36th international conference on machine learning. PMLR, pp 3448–3457. https://proceedings.mlr.press/v97/kleindessner19a.html
11. Kleindessner M, Awasthi P, Morgenstern J (2020) A notion of individual fairness for clustering. CoRR. https://arxiv.org/abs/2006.04960
12. Mahabadi S, Vakilian A (2020) Individual fairness for k-clustering. In: Proceedings of the 37th international conference on machine learning. PMLR, pp 6586–6596. https://proceedings.mlr.press/v119/mahabadi20a.html
13. Deepak P, Abraham SS (2020) Representativity fairness in clustering. In: 12th ACM conference on web science. Association for Computing Machinery, pp 202–211. https://doi.org/10.1145/3394231.3397910
14. Pedreshi D, Ruggieri S, Turini F (2008) Discrimination-aware data mining. In: Proceedings of the 14th ACM SIGKDD international conference on knowledge discovery and data mining. Association for Computing Machinery, pp 560–568. https://doi.org/10.1145/1401890.1401959
15. Rawls J (2001) Justice as fairness: a restatement. Principles of justice. The Belknap Press of Harvard University Press, Cambridge, MA, pp 39–79
16. Shah K, Gupta P, Deshpande A, Bhattacharyya C (2021) Rawlsian fair adaptation of deep learning classifiers. In: Proceedings of the 2021 AAAI/ACM conference on AI, ethics, and society. Association for Computing Machinery, pp 936–945. https://doi.org/10.1145/3461702.3462592

Hybrid Explainable Educational Recommender Using Self-attention and Knowledge-Based Systems for E-Learning in MOOC Platforms

Mehbooba P. Shareef⊙, Linda Rose Jimson, and Babita R. Jose

1 Introduction

Recommendation systems make product recommendations for users based on the in-depth user behaviour analysis carried out by the system in terms of previous purchase history of the user, the items users have added to cart/wish lists, ratings provided for different products, comments/feed backs on products, etc. Almost all the leading online service providers are using recommendation systems to increase their revenue and to help users make quick and correct choices from a subset of items. Generation of this subset of items with item ordering based on relevance of the item to the user or some other criteria is the responsibility of the recommendation system. There are lots of algorithms to make this possible. The most basic but efficient algorithm that is widely used for generating product/service recommendation is the collaborative filtering method in which recommendations are made based on preferences of the user under consideration and that of similar users with similar purchase history. A content-based recommendation engine makes recommendations based on the content/tags/metadata of the item to be recommended. Since every method has limitations, a hybrid approach always helps to combine the advantages and neutralise the draw-backs. Natural language processing on the reviews made by the user can also be proved instrumental in making correct recommendations since reviews also carry substantial information. Even though the development of the first recommendation engine dates back to as early as 1992 in an experimental mail sys-

M. P. Shareef · L. R. Jimson · B. R. Jose
Cochin University of Science and Technology, Kochi, India
e-mail: 19cs05lind@cusat.ac.in

B. R. Jose
e-mail: babitajose@cusat.ac.in

M. P. Shareef (✉)
Rajagiri School of Engineering and Technology, Kochi, India
e-mail: mehbooba@cusat.ac.in

© The Author(s), under exclusive license to Springer Nature Singapore Pte Ltd. 2022
J. Mathew et al. (eds.), *Responsible Data Science*, Lecture Notes
in Electrical Engineering 940, https://doi.org/10.1007/978-981-19-4453-6_4

Table 1 COCO data set: statistics

Learners	2,426,398
Courses	43,113
Learner ratings	4,530,508
Learner comments	1,266,313
Instructors	19,384
Average instructor ratings	300.74
Average enrollments per instructor	4036.39

Table 2 COCO data set: tables in data set

Table name	Field names
course	course_id, short_url, short_description, long_description, objectives, requirements, language, first_level_category, second_level_category, instructional_level, target_audience, subtitles
curriculum_lesson_chapter	resource_id, course_id, title, class, description, asset_type, asset_name, sort_order, created
evaluate	learner_id, course_id, learner_rating, learner_timestamp, learner_comment, review_id
instructor	instructor_id, job_title, total_reviews, total_enrollments, total_courses]
teach	course_id, instructor_id

tem known as tapestry, there are very few recommendation engines for educational recommendation. Since early 2020, due to the emergence of the pandemic named COVID-19, the educational sector has faced a paradigm shift to the online mode world over. Recommending students with the most appropriate learning materials is the essential part of online learning platforms. The data set used for our study aiming at bridging the aforementioned research gap is COCO [1]. COCO is undoubtedly the best data set that exists for technology enhanced learning with details about 43,000 courses, 16,000 instructors, 25 million courses, 45 million ratings and 12 million learners. The statistical details of COCO are given in Tables 1 and 2.

2   Related Works: Recommendation Systems in Education

An extensive study of the literature for research works on educational recommendation system is necessary to identify the potential research gaps in the area, and it can also reveal the major focus areas. We have considered more than 100 papers for the study and consolidated the methodology used in a representative sample of the papers.

Bourkoukou et al. [2] use user-similarity-based collaborative filtering followed by sequential pattern mining to advise automatic recommendations to an active learner based on his/her learning style, grades and user preferred learning material.

Klašnja-Milićević et al. [3] use collaborative tagging, tensor factorization and clustering methods to create a tag-based model for dynamic content recommendation.

Rahman et al. [4]use content-based recommendation and it augment web search engines with personalised recommendations of search results which match students' learning competencies and behaviours.

De Medio et al. [5] employ a hybrid system (content based + collaborative filtering) to recommend learning objects after sorting through supported standard list of learning object repositories (LOR).

Bhaskaran et al. [6] use clustering-based recommender, and it analyses user behavioural patterns using cluster-based pattern mining algorithms to understand learning patterns and recommends learning objects.

Ibrahim et al. [7] use fog-based recommendations for cloud. It employs fuzzy logic, ontology and association rule mining to get the best results for a query.

El Mabrouk et al. [8] use collaborative filtering, data mining techniques and user learner similarity metrics to classify items and make proper recommendation.

Niknam et al. [9] make use of fuzzy C-mean algorithm and ant colony optimization algorithms to groups learners based on their prior knowledge to make suitable recommendations.

Ghosh et al. [10] carry out query-lattice optimisation and makes recommendations based on association rule mining with reference to the query raised by the user.

Bhaskaran et al. [11] deploy a hybrid recommendation system using support vector machines, anarchic society optimisation and produces accurate recommendations for public data sets including learning recommendations for open university data recommendation systems in educational sets.

Khalid et al. [12] propose an online recommendation system using the concept of hyper spheres to cluster the data points. [12] is an incremental recommendation system which scales well with huge data sets. Their model is tested on COCO data set, and the model has very good coverage and lower error rate.

Zhang et al. [15] use deep belief networks to produce recommendations of resources in a MOOC environment. It uses learner features, attributes of course related to its content and behavioural patterns of learners. It performs supervised classification where the reviews given by learners act as the labels. They pre-train the model using unsupervised algorithms and use supervised learning to fine-tune the hyper-parameters and come out with a model which has faster convergence compared to traditional recommendation systems.

Hou et al. [16] propose an online course recommendation system which considers the sequential ordering of the learning curriculum. They use reinforcement learning to create a reward-based recommendation system which can adapt to changes faster and also the rewards are saved for future recommendations.

Pang et al. [17] propose a recommendation system to reduce the drop-out rate of learners in MOOC platforms. They predict the intensity of learning of a student with the help of Hawkes process and uses the knowledge distance to predict various knowledge features taking into account the learner-neighbours and learning series.

Zhao et al. [18] suggest a visual recommendation system which helps students to find out and play back helpful videos based on similarity with the videos that were watched by the learner previously and the relationship among different topics.

Zhang et al. [19] use distributed computations precisely association rule mining to figure out different rules which govern the choice of a course by a learner to give stable and accurate recommendations.

Wu et al. [20] combine collaborative filtering and the concept of metapaths to provide recommendations. Different metapaths will have different level of significance to different courses. By embedding metapaths and making use of attention mechanisms, they make the results interpretable to the user.

We have analysed the top 100 Google Scholar results for the search string 'recommendation systems', filtering the research works happened during 2015–2022 only and found that there are only 9% articles on E-learning-based recommendation systems, only 5% articles talks about explainability feature and that there are absolutely no research works on explainable educational recommendation systems. The analysis results are summarised in Fig. 1. This analysis shows the research gaps in educational recommendation systems, explainable recommendation systems and most importantly the scarcity of explainable educational recommendation systems which are the need of the hour. So in this work, we are trying to bridge this very evident research gap by proposing a novel explainable educational recommendation system using self-attention networks and knowledge-based systems.

3 Proposed System

Our proposed system predicts the rating of a course given the pair $\langle l, c \rangle$ as input where l is the learner_id and c is the course_id based on lc, the learner comment. To reach this goal our proposed method is equipped with two main modules (The architecture of the proposed method is given in Fig. 2). (1) A self- attention network which performs attention on learner-ID embeddings, course-ID embeddings and learner-comment embeddings to predict the ratings for learner-course pairs, (2) A knowledge-based system which considers average rating of a course, total number of times the course is offered and average rating of the instructor who is handling the course to find the relevance of the course to a learner. Both these modules go hand in hand to produce the final rating of a course as per the algorithm below.

3.1 Self-attention Module

The self-attention model has two neural networks running in parallel, one is for learner_modeling (Net_l) and another one for course_modeling(Net_c). The con-

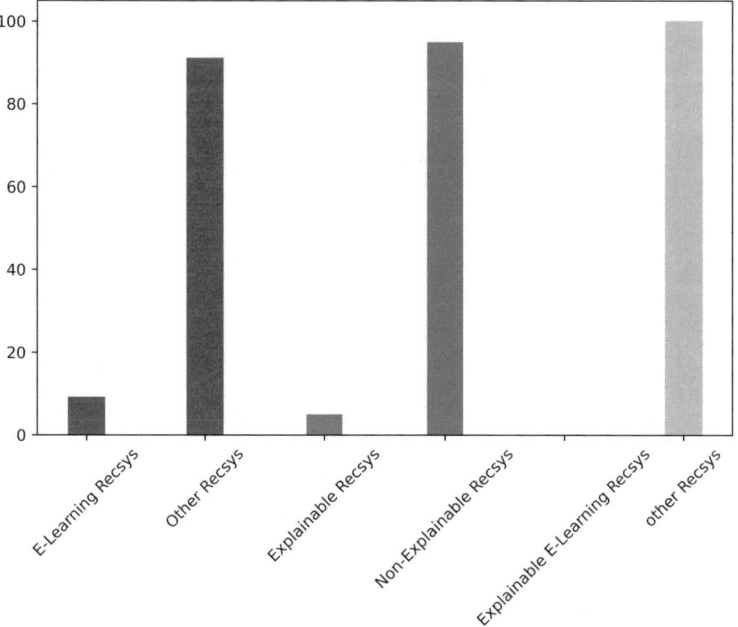

Fig. 1 Quantitative analysis of various recommendation systems

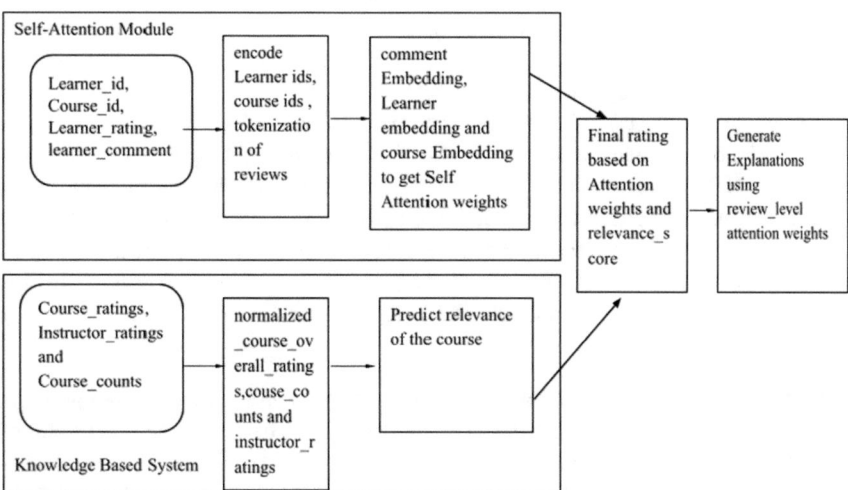

Fig. 2 Architecture of proposed system

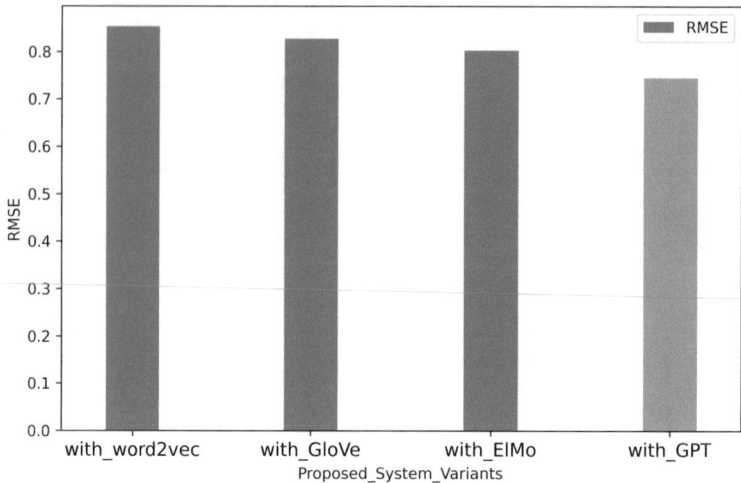

Fig. 3 Performance comparison of variants of proposed method

struction of this module is inspired by the work [12]. For review_embedding, we have analysed the performance of different embedding schemes like Word2Vec, GloVe, ELMo and GPT2.

- Word2Vec: We have used the pre-trained Word2Vec embedding model which was trained on Google News Data(GoogleNews-vectors-negative300.bin.gz). The number of embedding dimensions is 300.
- GloVe: We have used pre-trained GloVe embedding model which was trained on wikipedia data (glove.6B.300d.txt). The number of embedding dimensions is 300.
- ELMo: ELMo uses contextualised word embedding which uses information of all the words in a sentence to decide the embedding for a particular word. It is a pre-trained model, and the number of embedding dimensions is 1024.
- GPT2: We have used pre-trained model for GPT2 which uses absolute position embedding so as to give contextualised embeddings. The number of embedding dimensions is 768.

Comparing the RMSE of the model under all these embedding techniques, we could find that GPT2 performs the best and hence, we have used GPT2 to embed learner comments. The contextual encoding provided by GPT2 creates more meaningful review embedding is the reason. But since the number of embedding dimensions are comparatively higher, its time complexity is slightly higher than Word2Vec and GloVe but lesser than ELMo. The comparison results are shown in Figs. 3 and 4.

The self-attention module calculates learner features, course features, review features, learner level review attention, course level review attention and use these values to predict a rating as explained in the algorithm.

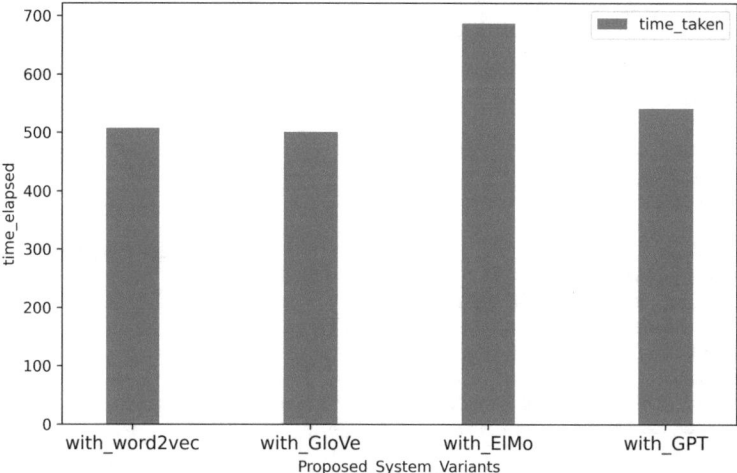

Fig. 4 Time-complexity analysis of variants of proposed method

3.2 Knowledge-Based System

The self-attention module takes into account only the learner_id, course_id, learner_rating and learner_comment. Apart from this, there are other features which are instrumental in deciding the relevance of a course to the learner. The knowledge-based system can extract relevant features by analysing the learner related, course related and instructor related information from the data set. The COCO data set contains lots of attributes of all these entities. Human expertise can be used to decide the features which will help in the process of predicting a course which is relevant to a learner. Thus, the knowledge-based system takes as inputs average rating of a course, total number of times the course is offered and average rating of the instructor who is handling the course and normalise these parameters to get normalised course average rating(CAR), normalised course count(CC) and normalised instructor performance(IP) so that all these values lie between [0–1]. And uses all these information to find the relevance score.

3.3 Explanation of the rating-prediction algorithm

- Steps 1–4 :Learner Ids need not be numbered from zero in real-life data sets. These steps extract unique learner_ids and number them from zero and transform all the learner_ids to this new encoding.
- Steps 5–8: Course Ids need not be numbered from zero in real-life data sets . These steps extract unique course_ids and number them from zero and transform all the course_ids to this new encoding.

Algorithm 1 Rating prediction

1: **for** $l \in L$ **do**   ▷ Embedding Learner ids
2:    $unique\_l \leftarrow L.index(l)$
3:    $l\_Feature \leftarrow embed\_id(l)$
4: **end for**
5: **for** $c \in C$ **do**   ▷ Embedding Course ids
6:    $unique\_c \leftarrow C.index(c)$
7:    $c\_Feature \leftarrow embed\_id(c)$
8: **end for**
9: **for** $lr \in LR$ **do**   ▷ Embedding Learner comments
10:    $review\_tokens \leftarrow tokenize(lc)$
11:    $embeded\_review\_tokens \leftarrow GPT2(tokens)$
12:    $lr\_Features \leftarrow BLH(embeded\_review\_tokens)$
13: **end for**
14: **for** $l \in L$ **do**   ▷ Finding Learner level review attention
15:    $learner\_review\_attention \leftarrow review\_attention(l\_Feature, lr\_Features)$
16:    $learner\_review\_attention \leftarrow learner\_review\_attention + learner\_review\_attention *$
    $lr\_Features$
17: **end for**
18: **for** $c \in C$ **do**   ▷ Finding Course level review attention
19:    $course\_review\_attention \leftarrow review\_attention(c\_Feature, lr\_Features)$
20:    $course\_review\_attention \leftarrow course\_review\_attention + course\_review\_attention *$
    $lr\_Features$
21: **end for**
22: **for** $c \in C$ **do**   ▷ Finding Course relevance
23:    $courseval \leftarrow avg(Course\_Average\_Rating, Course\_Count, Instructor\_Performance)$

24:    $rel \leftarrow \begin{cases} VeryHigh & \text{if courseval} >= \text{High} \\ High & \text{if courseval} >= \text{Medium} \\ Medium & \text{if courseval} > \text{Low} \\ Low & \text{if courseval} = \text{Low} \\ VeryLow & \text{otherwise} \end{cases}$

25: **end for**
26: $rating \leftarrow aggregate(learner\_review\_attention, l\_features, course\_review\_attention, c\_$
    $features, rel)$

- Steps 9–13 : These steps tokenize the already pre-processed reviews into words. The words are then embedded using a pre-trained GPT2 model. Learner_review features are then extracted with the help of a bidirectional LSTM with all the hidden layer nodes as outputs [13].
- Steps 14–17: Apply self-attention on learner ID features and corresponding learner's learner comment features. This learner review attention is multiplied with learner review features and is added to existing learner review attention for updating the latter.
- Steps 18–21: Apply self-attention on course ID features and corresponding course's course comment features. This course review attention is multiplied with course review features and is added to existing course review attention for updating the latter.
- Steps 22–25: For each course, a relevance score is calculated.

Table 3 RMSE comparison with baselines

Method	RMSE
NARRE_FA	0.984114
NARRE_SOA	0.955944
NARRE_LSTM	0.766035
NARRE_LSTM_SA	0.818802
NARRE_CNN_LSTM	0.778785
NARRE_CNN_LSTM_SA	0.811133
HANCI	0.835487
Proposed method	
with_word2vec	0.853408
with_GloVe	0.827432
with_ELMo	0.803274
with_GPT	**0.745574**

- Step 26: Total rating of the course is calculated as a function of learner review attention, learner features, course review attention, course features and the relevance score.

4 Experimental Results

We have analysed the performance of the model on a subset of COCO data set and compared its performance with the variants of [13] and [14] since our method is a modified model of the former and found that our model outperforms the baseline architectures. Root mean square error (RMSE) is the evaluation metric that we have used.

$$\text{RMSE} = \sqrt{\frac{1}{N} \Sigma_{l,c} (\hat{R}_{l,c} - R_{l,c})^2}$$

$\hat{R}_{l,c}$ is the predicted rating for learner l on course c and $R_{l,c}$ is the actual rating for the learner l on course c. The results are consolidated in Tables 3 and Fig 5.

We have also compared our model with the models implemented on COCO data set. Asra Khalid et al. [12] implemented an online recommendation system for MOOC named NoR-MOOCs and created other implementations using K-means clustering, collaborative filtering-based recommendation system and a random recommendation system. All the models were tested on COCO, and they have recorded the RMSE after testing. Our algorithm outperforms all the recommendation system variants implemented in [12] by a good margin, and the results are shown in Table 4 and Fig 6.

M. P. Shareef et al.

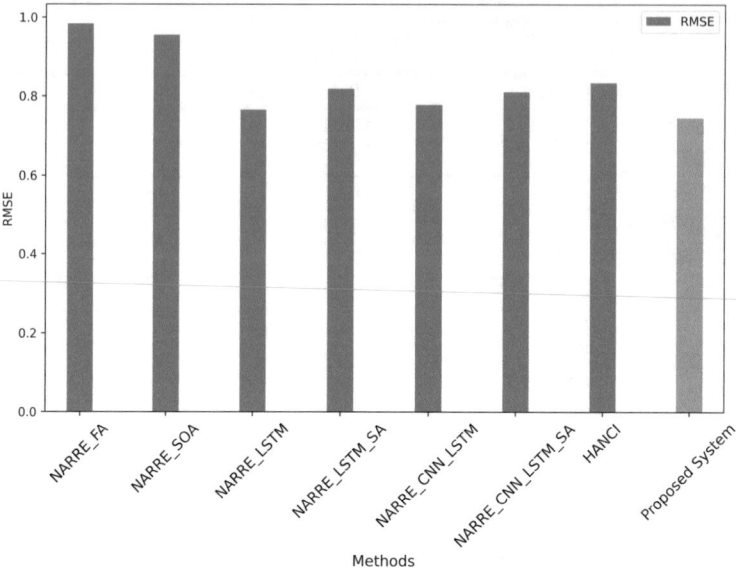

Fig. 5 Comparison with baseline methods

Table 4 RMSE comparison with E-learning recommendation systems tested on COCO

Method	RMSE
K-means	0.8359
Collaborative filtering	0.8374
K-Means	0.766035
Random	0.8123
NoR-MOOCs	0.7908
Proposed method	**0.745574**

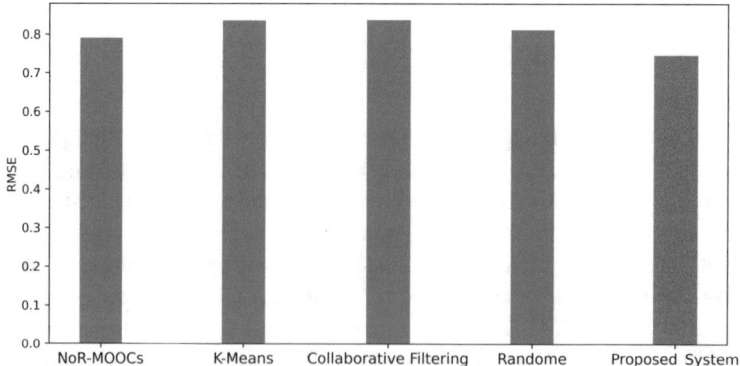

Fig. 6 Comparison with E-learning recommendation systems tested on COCO

Table 5 Review with highest attention weight

Attention weight	Review
0.005849681329	Just INCREDIBLE! Like, honestly, Anette makes it SO simple to understand AND she gives us great examples which anyone can relate to. Really a Blessing! Thank You.

Table 6 Review with lower attention weight

Attention weight	Review
0.005846943706	Love the information on the reframing, and benefit finding, and the Robinson Caruso method. I can use with clients, myself, and know someone personally I'm going to refer to this course

4.1 Explainability Analysis

Explainability of the model can be inferred from the review level and word level attention weights. For example, two reviews with different attention weights are compared. High attention weights of reviews indicate usefulness of a review to the learner in choosing a course and attention weights of words in a review indicate how instrumental those words were in making the review useful to a learner. The reviews were pre-processed to remove stop words from and the tokenized reviews were embedded using pre-trained GPT2 model which provides contextualised embedding. Features of reviews were extracted using bidirectional LSTM with all the hidden layers as outputs. The word level attention weights of reviews were obtained by interacting the features of learners and the aforementioned review features. The explainability of the model is analysed in this session. Table 5 shows the review with highest attention weights, and Table 6 shows another review having lower attention weights. Table 7 shows the word level attention weights of both reviews, and Fig 7 graphically represents the same for easier understanding

The review with higher attention weight is undoubtedly more helpful to the learner since it is having words like incredible, example, simple and blessing which are all useful adjectives when it comes to a course or instructor. The review with lower weight is missing such adjectives which are very catchy from a learner's perspective. Hence, an explanation can be provided to the user with the help of the attention weights assigned to different words in the review. A generative adversarial neural network model can be used here to generate a proper explanation from the words having highest attention weights considering all the potential reviews of the item which has been recommended. Proper explanations for the recommendations help a learner to choose the next course wisely from a ranked list of recommendations.

Table 7 Word-level attention weights

Words	Attention
'incredible', 'honestly', 'anette', 'simple', 'understand', 'examples', 'relate', 'blessing'	−0.24458308517932892, −0.2795969545841217, −0.3059690296649933, −0.319175660610199, −0.32151395082473755, −0.3163332939147949, −0.2971207797527313, −0.2703518271446228
'love', 'reframing', 'benefit', 'finding', 'robinson', 'caruso', 'method', 'clients'	−0.25301918387413025, −0.2873460352420807, −0.3171443045139313, −0.33144640922546387, −0.3369181156158447, −0.3259398937225342, −0.3046564757823944, −0.2793954610824585

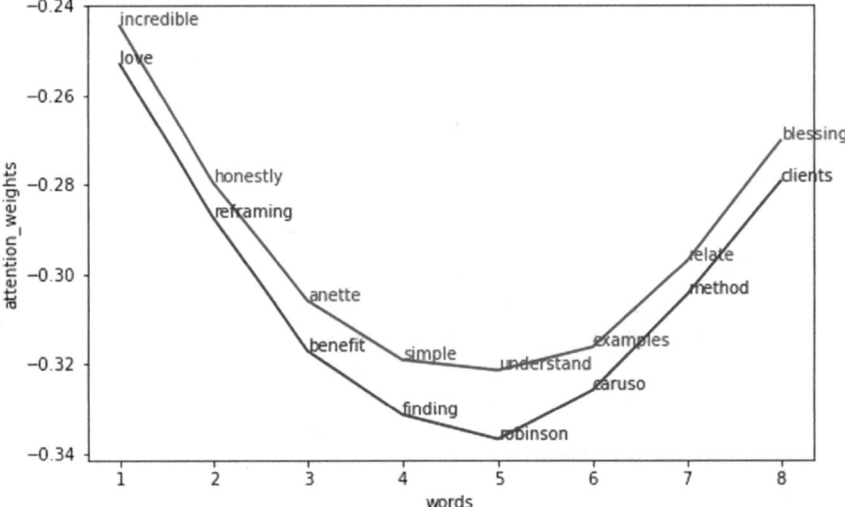

Fig. 7 Word-level attention weights

5 Conclusion and Future Work

There are very few works in the literature on educational recommendation systems. In a random sample of 100 research papers taken from the period 2015-2022, we could not see any work on explainable educational recommendation system. When education systems are undergoing a paradigm shift to online platforms (technology enhanced learning) , we find this as a potential research gap. Our model produces educational recommendations for MOOC platforms employing self-attention and knowledge-based systems. This hybrid system uses learner level attention, course level attention and learner's review level attention and makes use of a knowledge-based system to extract content aware information to sharpen the predictions of the network, thereby lowering RMSE by a considerable amount. We have used GPT2 to create contextualised embedding of learner comments. The experimental results show that it outperforms the baseline architectures we have compared it with, both educational recommenders [12] and e-commerce recommenders [13],[14].

We will be extending this work including GANs to output the explanations to learners in a better format since explanations for recommendations are really important in an online learning platform where a student will be really wanting to know the rationale behind the suggestions to decide on the next step. We are looking into making this recommendation system adaptive and incremental so that it can respond faster to changes in user preferences with time, can recommend newly added courses without any cold start issues. We are also looking into make use of mathematical models that efficiently employs transitional probability so that the learning and course sequence information can be processed much better. This is due to the intuition that in course recommendations, it is not the similarity among different courses that must be given priority, but the sequential relationship among different learning objects.

References

1. Dessì D, Fenu G, Marras M, Recupero DR (2018) Coco: semantic-enriched collection of online courses at scale with experimental use cases. In: World conference on information systems and technologies. Springer, Cham, pp 1386-1396
2. Bourkoukou O, El Bachari E, El Adnani M (2017) A recommender model in e-learning environment. Arab J Sci Eng 42(2):607–617
3. Klašnja-Milićević A, Ivanović M, Vesin B, Budimac Z (2018) Enhancing E-learning systems with personalized recommendation based on collaborative tagging techniques. Appl Intell 48(6):1519–1535
4. Rahman MM, Abdullah NA (2018) A personalized group-based recommendation approach for web search in E-learning. IEEE Access 6:34166–34178
5. De Medio C, Limongelli C, Sciarrone F, Temperini M (2020) MoodleREC: a recommendation system for creating courses using the moodle E-learning platform. Comput Hum Behav 104:106168
6. Bhaskaran S, Marappan R, Santhi B (2021) Design and analysis of a cluster-based intelligent hybrid recommendation system for E-learning applications. Mathematics 9(2):197
7. Ibrahim TS, Saleh AI, Elgaml N, Abdelsalam MM (2020) A fog based recommendation system for promoting the performance of E-Learning environments. Comput Electr Eng 87:106791
8. El Mabrouk M, Gaou S, Rtili MK (2017) Towards an intelligent hybrid recommendation system for E-learning platforms using data mining. Int J Emerg Technol Learn 12(6)
9. Niknam M, Thulasiraman P (2020) LPR: A bio-inspired intelligent learning path recommendation system based on meaningful learning theory. Educ Inform Technol 25(5):3797–3819
10. Ghosh S, Roy S, Sen S (2021) An efficient recommendation system on E-learning platform by query lattice optimization. In: Data management, analytics and innovation. Springer, Singapore, pp 73–86
11. Bhaskaran S, Marappan R (2021) Design and analysis of an efficient machine learning based hybrid recommendation system with enhanced density-based spatial clustering for digital E-learning applications. Complex Intell Syst 1–17
12. Khalid A, Lundqvist K, Yates A, Ghzanfar MA (2021) Novel online recommendation algorithm for massive open online courses (NoR-MOOCs). PLOS One 16(1):e0245485
13. Yang C, Zhou W, Wang Z, Jiang B, Li D, Shen H (2021) Accurate and explainable recommendation via hierarchical attention network oriented towards crowd intelligence. Knowl Based Syst 213:106687

14. Chen C, Zhang M, Liu Y, Ma S (2018) Neural attentional rating regression with review-level explanations. In: Proceedings of the 2018 world wide web conference, pp 1583–1592
15. Zhang H, Huang T, Lv Z, Liu S, Yang H (2019) MOOCRC: a highly accurate resource recommendation model for use in MOOC environments. Mob Networks Appl 24(1):34–46
16. Hou Y, Zhou P, Xu J, Wu DO (2018) Course recommendation of MOOC with big data support: a contextual online learning approach. In: IEEE INFOCOM 2018—IEEE Conference on computer communications workshops (INFOCOM WKSHPS), pp 106–111. IEEE
17. Pang Y, Liao C, Tan W, Wu Y, Zhou C (2018) Recommendation for MOOC with learner neighbors and learning series. In: International conference on web information systems engineering. Springer, Cham, pp 379–394
18. Zhao J, Bhatt C, Cooper M, Shamma DA (2018) Flexible learning with semantic visual exploration and sequence-based recommendation of MOOC videos. In: Proceedings of the 2018 CHI conference on human factors in computing systems, pp 1–13
19. Zhang H, Huang T, Lv Z, Liu S, Zhou Z (2018) MCRS: a course recommendation system for MOOCs. Multimedia Tools Appl 77(6):7051–7069
20. Wu L (2021) Collaborative filtering recommendation algorithm for MOOC resources based on deep learning. Complexity

An Improved Recommendation System with Aspect-Based Sentiment Analysis

Seema Safar, Babita R. Jose, and T. Santhanakrishnan

1 Introduction

Recommendation systems are an automated system capable of filtering any items such as products, restaurants, advertisements, people, movies or songs. We see this all over the world on a daily basis such as Amazon, Netflix, Pandora, and YouTube. For example, if we watch a movie, we get a recommendation on a different movie based on the power of previous viewing history. Recommendation system not only works in what products to be shown but also in what order the products are being ranked. At first, a non-personalized recommendation algorithm that recommends only the most popular products based on ranks was introduced. Popularity of a product can be determined by measuring its sales, ratings, and reviews. It is assumed that more popular products are likely to be preferred more by the users. The popularity-based approach is based on this concept [1]. However, the limitation of this approach is there is a chance of listing a not-so-preferred list of different products also be to the user [2].

Recommendations are responsible for 70% of the time people spend watching videos on YouTube, and 35% of the purchases on Amazon is the result of Amazon recommendation engine. Not all aspects contribute to improving the experience of the user with the item. Depending on the scenario, the importance of the aspect may vary. For example, in the case of the review of a restaurant, aspects may be the service provided, ambience of the place or the dishes in the restaurant, whereas in a movie, the aspect may vary.

S. Safar (✉) · B. R. Jose
School of Engineering, Cochin University of Science and Technology, Kochi, Kerala, India
e-mail: seemasafar@cusat.ac.in

B. R. Jose
e-mail: babitajose@cusat.ac.in

T. Santhanakrishnan
Naval Physical Oceanographic Laboratory, Defense Research and Development Organisation, Kochi, Kerala, India

© The Author(s), under exclusive license to Springer Nature Singapore Pte Ltd. 2022
J. Mathew et al. (eds.), *Responsible Data Science*, Lecture Notes
in Electrical Engineering 940, https://doi.org/10.1007/978-981-19-4453-6_5

Research studies in the neural organization-based aspect-based sentiment classification models have profoundly reshaped the research focus on a new dimension of predictions for recommendations [3]. In all these studies, context words are given adequate importance as that of the mentioned aspects. However, some works rely only on a segment of the context words to understand the polarity of the review as in [4].

Neural networks leveraging attention mechanisms were explored to differentiate important and relevant aspects from the user reviews [5, 6].

2 Recent Works

This section includes a brief review on the core areas highly relevant to the proposed work. The areas focused in this proposed work include: (1) Review-aware recommendation systems (2) Aspect-based sentiment analysis.

2.1 Review-Aware Recommendation Systems

Many studies have incorporated aspects of user reviews with ratings in order to improve recommendation performance. Reference [7] proposes an attentive aspect-based recommendation model (AARM) to model the interactions between synonymous and similar aspects. To learn multifaceted user preferences, traditional topic models were used in recommendation systems to extract topics, aspects, and features from reviews [8]. Though models such as TransNets [9] and D-Attn [10] have a significant prediction capability, the learned low-dimensional latent representations suffer from coarse-grained user preference extraction. Another variant of aspect-based recommendations, such as EFM [11] and LRPPM [12], relies on natural language processing tools for the extraction of aspects annotation tools are used to extract aspects and sentiments from reviews. Similarly, TriRank [13] constructed the user-item-aspect tripartite graph using the extracted aspect for recommendations. The contributions of some previous works under review-aware recommender systems relevant for our work, which include Deep CoNN [14], ANR [15], ANCF [16], and ALFM [17], are listed in Table 1.

2.2 Aspect-Based Sentiment Analysis

Aspect-based sentiment classification focuses on determining the sentiment polarity of a sentence by considering the aspect of the sentence [18, 19]. Recent research has turned its focus on developing end-to-end neural network models. Some of the previous works relevant for our work [6, 20–23] are listed in Table 1.

Table 1 Previous works

Category	Model	Backbone	Evaluation parameters P R MSE	Contribution
Review-aware recommendation systems	DeepCoNN	CNN	X X	Incorporates two deep neural networks coupled together to learn users and items from the review to enhance the predicting ratings
	TransNets	CNN	X X	Extended work of DeepCoNN, by including one more additional latent layer that represents the target user-target item pair
	ALFM	CNN	X X	Estimation of aspect ratings and assigning weights to different aspects
	AARM	Attention Network		Sparsity problem, heterogenity in aspect interest of the user with respect to different products is considered
	ANR	Attention Network	X X	Aspect-aware representation learning and an aspect importance estimator based on attention and co-attention is considered
	ANCF	Attention Network	X X	The problem of heterogenity in the aspect interest of the user is considered with a neural attention network

(continued)

Table 1 (continued)

Category	Model	Backbone	Evaluation parameters P R MSE	Contribution
Aspect-based sentiment analysis	Modeling inter-aspect	RNN	X X X	Inter-aspect dependencies are incorporated to improve the task of prediction
	IAN	LSTM	X X X	Attentions in the contexts and targets are learned to improve sentiment analysis
	ATAE-LSTM	LSTM	X X X	Learns aspect embeddings and uses aspects in computing the attention weights to improve aspect level sentiment analysis
	GRACE	Multi-head Attention	X X	Use a cascaded labeling approach to enhance the interaction between aspect terms by a multi-head attention architecture
	Hierarchial model	Hierarchial BiLSTM	X X X	Makes use of the structure of the review and the sentential context to make predictions

3 Proposed Methodology

The proposed model focuses on rating prediction from a user-item space by considering the aspect-based sentiment on the review provided by the user. The motivation of the proposed methodology is to improve the traditional attention-based recommendation system with aspect-based sentiment analysis. Instead of following the usual paradigm of approximating the attention matrix in attention layers here, the attention sublayers are completely replaced with a linear mixing, namely the fast Fourier transform [24]. This offers two important benefits:

(i) Model size: As there are no learnable weights in the embedding mixing layer, the number of model parameters has decreased drastically, thereby improving the training time

(ii) Training Time: Fast Fourier transform is used as the core support of the model. Fast Fourier transform is based on divide and conquer policy, and therefore, the time complexity of the method is $O(nlogn)$.

First, we present the overview of our model. This is followed by the description of the two main components of the model that helps in rating prediction.

The major contributions of this paper include: (i) An improved recommendation system with aspect-based sentiment analysis, by replacing the attention layer with the Fourier transform is developed. (ii) The proposed model is evaluated on standard laptop, restaurant datasets SemEval 2014 [19] and the manually annotated Twitter dataset introduced by Li et al. [3]. Experimental results show that using fast Fourier transforms offer a trade-off between efficiency and accuracy on the recommendation task.

3.1 Model Overview

The overall architecture of the proposed model includes two important components, one for predicting the polarity-based aspect-specific sentiment and the other for predicting the rating of the item from the sentiment polarity of the aspect-based sentiment prediction. The proposed framework of the model is shown in Fig. 1.

The raw data are preprocessed to build the vocabulary. For each word, word embedding is applied. A bidirectional GRU is applied on the word embedding to get a final word representation. Since aspect-based sentiment analysis is to be performed, we build an aspect embedding from the vocabulary to feed into the input layer of the

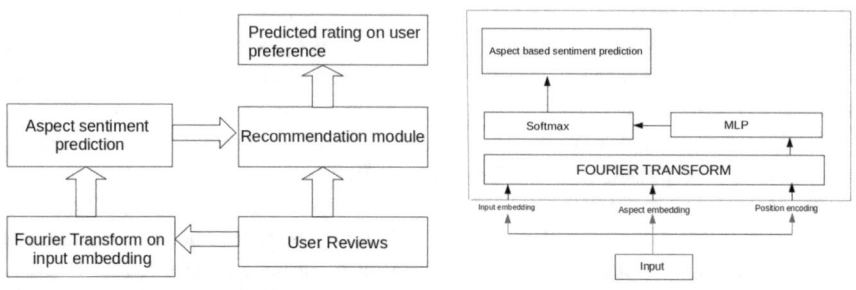

(a) The Proposed framework of the model   (b) Proposed Architecture of ABSA

Fig. 1 Proposed methodology

network. The embedding includes the input embedding and aspect embedding with the positional encoding incorporated in it. Each sequence of words can be viewed as a series of vectors, which is dependent on the number of tokens in the sentence.

A Fourier transform sublayer, which applies a 2D Fourier transform (typically Fast Fourier transform) to the resultant input embedding vector along the sequence length and the hidden dimension, is then applied, which results in different harmonics to analyze the various important aspects in a sentence and no important aspect would be left neglected. Since the word sequences are positionally encoded as shown in Fig. 1b, mixing the tokens while performing Fourier Transform does not affect the sequence of the word.

The transformed input embedding is then connected to the aspect sentiment prediction layer. The output of this module includes aspect sentiment distribution over three classes and aspect-specific sentence representations. The three classes considered here include (0: neutral, 1: positive, -1: negative). This layer is fed into the recommendation part of the proposed work, to predict the ratings of any user for instance 'u_i' on any item 't'.

3.2 Aspect-Based Sentiment Analysis

The proposed architecture of the aspect-based sentiment analysis (ABSA) is shown in the Figure 1b. As mentioned in Sect. 3.1, the input embeddings are fed into the aspect-based sentiment analysis part. A fast Fourier transform sublayer, which applies a two-dimensional fast Fourier transform to input embeddings (*FFTs*) and (*FFTh*), is applied along the embeddings, motivated from the work of [24].

$$y = \Re(\text{FFTs}(\text{FFTh}(x)))\qquad(1)$$

As indicated by Eq. (1), only, the real part of the result is considered, and the complex numbers are excluded, so as to keep the (nonlinear) feed-forward sublayers or output layers unmodified. The proposed model (FT_RECO) showed best results, when extracted only the real part of the transformation after applying FFTs and FFTh. The multidimensional vector is then passed through a multi-layer perceptron (MLP), which employs the gradient descent approach to ensure that the learning weights converge quickly. The vector is then passed through a Softmax activation function to compute the probability distribution over aspects.

Assume that there are 'n' aspects, the probability distribution over 'n' aspects can represent the possibility that the sentence belongs to a particular aspect. The review embedding is concatenated with the extracted aspect embedding along with the position encoding, which is passed through the feed-forward network. This fuses the information of aspects to the representation of target sentences, which is fed into an aspect-based sentiment prediction part. A Leaky ReLU activation function is applied in this layer to predict the aspect sentiment distribution on review 's' with respect to the aspect. Along with this, a probability of polarity score of a review's' *s*' with respect to each aspect is also computed.

3.3 Recommendation Module

This module predicts the rating of user toward the item. The input to this module will include the matrix with the information of user-item interaction. Following the work of [25], to capture the semantic information, a convolution neural network is employed. This network performs convolutions on each matrix of reviews. This step is followed by a max pooling, on a subset of reviews. This helps to find out the most important features in the sentence. The resultant matrix is formed by concatenating the outputs with same aspects. The aspect importance of each user with respect to the item is inferred from the user-rating matrix and the review embedding. The sentiment polarity score from the aspect-based sentiment prediction part is also taken into consideration, to finalize the value of aspect importance which outputs whether the item is recommendable or not. Based on this, the recommendation module predicts the rating of the item.

4 Experiments

To evaluate the proposed model, extensive experiments were conducted to answer the following research questions (RQ1, RQ2, RQ3).

1. RQ1: How does the proposed aspect-based sentiment module perform compared to the state-of-the-art existing aspect-based sentiment analysis models used for aspect-based sentiment analysis? (Sect. 4.3)
2. RQ2: How does the recommendation module performs in terms of performance compared to the current state-of-the-art models? (Sect. 4.3)
3. RQ3: How does the improved approach contributes to the overall performance? (Sect. 4.3).

Using an embedding layer, each word in the vocabulary is transformed into its corresponding '*n*' dimensional vector, according to the user reviews. In order to represent user reviews and aspects in a better manner, the embedding matrix is built with word vectors that have been pretrained on large corpora, such as GloVe3 [26]. The order of words is preserved in the embedded vector. The dimension of word vectors, aspect embeddings, and the size of hidden layer are 300.

To evaluate the performance of the proposed model, we have split the dataset into 80:20 ratio. The dropout value of the model was selected as 0.2 with the batch size of 32. The model was trained on 50 epochs, with a learning rate 0.001 and the optimizer as Adam.

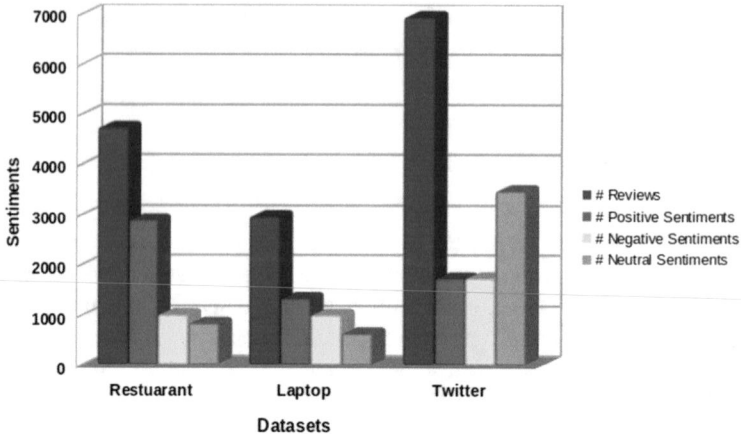

Fig. 2 Statistics of dataset

4.1 Dataset

The proposed model (FT_RECO) is experimented on three publicly available datasets, which include the reviews of restaurant and laptop domains from SemEval 2014 [19] and manually annotated Twitter dataset collected by Li et al. [3]. The statistics of the dataset are shown in Fig. 2. The conflict category in SemEval 2014 datasets is removed by balancing the datasets.

4.2 Baseline

Exploratory analysis on the following baseline models was performed for the analysis of the aspect-based sentiment analysis(ABSA) and recommendation. To ensure RQ1 and RQ2 in Sect. 4, the evaluation of each of the units : Aspect-based sentiment analysis, recommendation module are shown in Fig. 3 and Table 5.

The baseline models used for the analysis of aspect-based sentiment analysis and recommendation include:

AT-LSTM [23]: In this model, the important features or the aspects of the review were extracted making use of attention mechanism.

ATAE-LSTM [23]: In this model, each input word vector is affixed with the aspect embedding. The final representation is generated by leveraging an attention mechanism.

MEMNET [27]: In this model, a multi-hop attention mechanism is employed on the memory, and the memory is stacked by input word embedding. An attention mechanism is incorporated at a previous hop so as to improve the distribution of attention at a later hop.

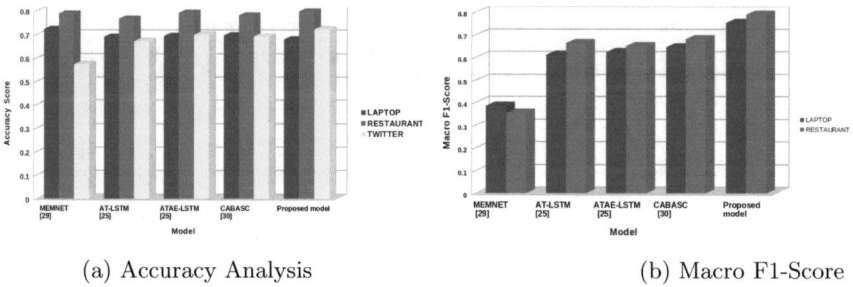

(a) Accuracy Analysis   (b) Macro F1-Score

Fig. 3 Performance analysis of aspect-based sentiment analysis

CABASC [28]: In this model, a content mechanism at sentence level and a context level attention mechanism to solve sentiment mismatch problem is also employed.

MRM [25]: In this model, an interactive attention mechanism is employed for recommending user predictions.

4.3 Performance Analysis

The model performance in aspect-based sentiment analysis part of the proposed model(FT_RECO) with the baseline models is analyzed and is shown in Fig. 3. The evaluation metrics used include: accuracy, macro F1-score.

To ensure RQ2 (Sect. 4), an exploratory analysis of the improved recommendation system with aspect-based sentiment analysis was performed by comparing with the baseline model [23], by fine-tuning the hyperparameters in the proposed work. In the first environment, the word embedding vector and the aspect embedding vector values were fine-tuned. The second environment was set up by setting the word embedding vector fixed and the fine-tuning the aspect embedding vector. The third environment was set up in such a way, that the aspect embedding vector was fixed, and the word embedding vector was fine-tuned, to understand how the parameter tuning influenced the performance of the recommendation systems. The performance of the proposed model (FT_RECO) with respect to each environment is shown in Tables 2, 3 and 4. The same environment was set up in the baseline model [23], being the successful baseline model, which incorporated both aspect embedding as well as attention mechanism in it. The performance comparison of both [23] and the proposed model (FT_RECO) done on the basis of the evaluation metric [(Precision, $F1$-Score, Root Mean Square Error (RMSE)] is shown in Tables 2, 3, and 5.

The time taken for training the proposed model (FT_RECO) is compared with the traditional attention-based models to understand the impact of Fourier transform in the proposed model. Figure 4 depicts that the accuracy of proposed model (FT_RECO) converges quickly without much oscillations, compared to attention-based models such as ATAE-LSTM [23], AT-LSTM [23], CABASC [28]. This makes it evident

Table 2 Environment 1: fine-tuning word embedding and aspect embedding

Dataset	Evaluation metric	ATAE-LSTM [23]	FT_RECO [proposed]
Laptop	Precision	0.67	**0.69**
	F1-score	0.62	**0.63**
	RMSE	0.78	**0.76**
Restaurant	Precision	0.75	**0.76**
	F1-score	0.64	**0.75**
	RMSE	0.69	**0.67**
Twitter	Precision	0.69	**0.69**
	F1-score	0.68	**0.70**
	RMSE	0.64	**0.63**

Table 3 Environment 2: Fine-tuning aspect embedding

Dataset	Evaluation metric	ATAE-LSTM [23]	FT_RECO [proposed]
Laptop	Precision	0.67	**0.683**
	F1-score	0.61	**0.626**
	RMSE	0.73	0.767
Restaurant	Precision	0.74	**0.767**
	F1-score	0.69	**0.698**
	RMSE	0.683	**0.624**
Twitter	Precision	0.683	**0.699**
	F1-score	0.673	**0.677**
	RMSE	0.643	**0.641**

Table 4 Environment 3: Fine-tuning the word embedding

Dataset	Evaluation metric	ATAE-LSTM [23]	FT_RECO [proposed]
Laptop	Precision	0.72	**0.72**
	F1-score	0.64	**0.64**
	RMSE	0.75	**0.73**
Restaurant	Precision	0.76	**0.777**
	F1-score	0.69	**0.671**
	RMSE	0.641	**0.666**
Twitter	Precision	0.677	**0.699**
	F1-score	0.675	**0.684**
	RMSE	0.644	**0.626**

that there are no learnable parameters, since there are no heavily trainable parameters as that of attention layer in the proposed model.

Table 5 shows the overall performance of the proposed model in English reviews in comparison with [25] (MRM), which uses attention layer for multilingual languages

Table 5 Recommendation module analysis

	Model	Precision	Recall	MSE
Environment 1	FT_RECO [proposed]	**0.76**	**0.76**	**1.091**
	MRM [25]	0.77	0.773	1.111
Environment 2	FT_RECO [proposed]	**0.76**	**0.76**	**1.151**
	MRM [25]	0.75	0.75	1.214
Environment 3	FT_RECO [proposed]	**0.78**	**0.78**	**1.142**
	MRM [25]	0.78	0.773	1.153

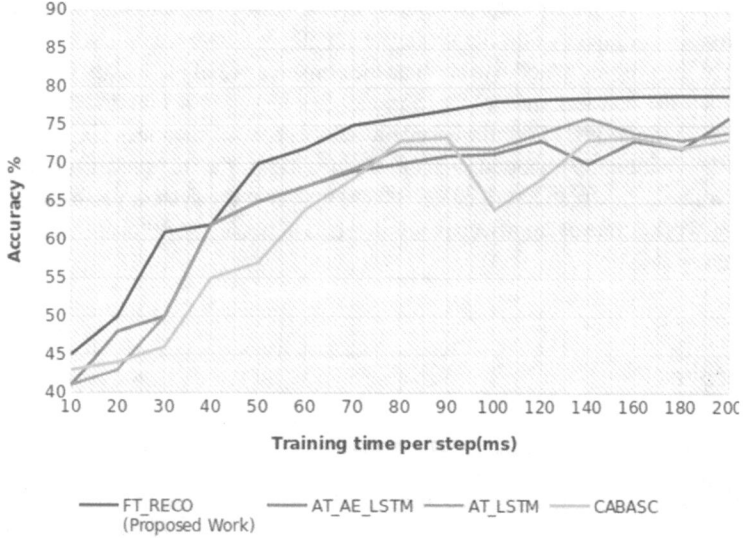

Fig. 4 Training time per step versus accuracy of the proposed model (FT_RECO) in comparison with other attention-based models

like English and French. The proposed model is trained on English reviews only. Performance analysis of recommendation module with the state-of-the-art model [25] (MRM) for restaurant reviews in English language on all three fine-tuned environment is shown in Table 5. Precision, recall, and mean square error (MSE) are the evaluation metrics used for evaluating the recommendation. The bold numbers show the improvements in the performance of the proposed model, which ensures RQ3 (Sect. 4).

5 Conclusion

The proposed model presents a novel approach for the high-demand recommendation task. In this work, a simplified improved recommendation system with aspect-based sentiment analysis using fast Fourier transform is studied. The contributions of the work include

We show that simple, linear token (input embeddings) mixing transformations, along with the nonlinearities in feed-forward layers are sufficient enough to model different semantic relationships in text. An improved recommendation system with aspect-based sentiment analysis using fast Fourier transform (FT_RECO) is introduced, where the high-parameterized attention sublayer is replaced by an fast Fourier transform. The simple equation in the paper clearly shows that the proposed model uses less number of parameters. Since the training time of any model is dependent on the learnable parameters and the sequence length. We investigate the training time per step of the proposed model to analyze the convergence of the weights in the model to establish the fact that the proposed model has less parameters and hence simple compared to traditional attention-based models. The word sequences in the proposed model are positionally encoded to avoid the mixing of the tokens while performing Fourier transform, which has does not adversely affect the system performance. The proposed work achieves improved result in terms of accuracy, precision, recall, mean square error (MSE).

References

1. Ricci F, Rokach L, Shapira B (2015) Recommender systems: introduction and challenges. In: Recommender systems handbook. Springer, Berlin, pp 1–34
2. Cremonesi P, Koren Y, Turrin R (2010) Performance of recommender algorithms on top-n recommendation tasks. In: Proceedings of the fourth ACM conference on recommender systems, pp 39–46
3. Dong L, Wei F, Tan C, Tang D, Zhou M, Xu K (2014) Adaptive recursive neural network for target-dependent twitter sentiment classification. In: Proceedings of the 52nd annual meeting of the Association for Computational Linguistics (volume 2: Short papers), pp 49–54
4. Tang D, Qin B, Liu T (2016) Aspect level sentiment classification with deep memory network. *arXiv preprint* arXiv:1605.08900
5. Chen P, Sun Z, Bing L, Yang W (2017) Recurrent attention network on memory for aspect sentiment analysis. In: Proceedings of the 2017 conference on empirical methods in natural language processing, pp 452–461
6. Ma D, Li S, Zhang X, Wang H (2017) Interactive attention networks for aspect-level sentiment classification. *arXiv preprint* arXiv:1709.00893
7. Guan X, Cheng Z, He X, Zhang Y, Zhu Z, Peng Q, Chua T-S (2019) Attentive aspect modeling for review-aware recommendation. ACM Trans Inform Syst (TOIS) 37(3):1–27
8. Diao Q, Qiu M, Wu C-Y, Smola AJ, Jiang J, Wang C (2014) Jointly modeling aspects, ratings and sentiments for movie recommendation (JMARS). In: Proceedings of the 20th ACM SIGKDD international conference on knowledge discovery and data mining, pp 193–202
9. Catherine R, Cohen W (2017) Transnets: learning to transform for recommendation. In: Proceedings of the eleventh ACM conference on recommender systems, pp 288–296

10. Seo S, Huang J, Yang H, Liu Y (2017) Interpretable convolutional neural networks with dual local and global attention for review rating prediction. In: Proceedings of the eleventh ACM conference on recommender systems, pp 297–305
11. Zhang Y, Lai G, Zhang M, Zhang Y, Liu Y, Ma S (2014) Explicit factor models for explainable recommendation based on phrase-level sentiment analysis. In: Proceedings of the 37th international ACM SIGIR conference on research & development in information retrieval, pp 83–92
12. Chen X, Qin Z, Zhang Y, Xu T (2016) Learning to rank features for recommendation over multiple categories. In: Proceedings of the 39th international ACM SIGIR conference on research and development in information retrieval, pp 305–314
13. He X, Chen T, Kan M-Y, Chen X (2015) Trirank: review-aware explainable recommendation by modeling aspects. In: Proceedings of the 24th ACM international conference on information and knowledge management, pp 1661–1670
14. Zheng L, Noroozi V, Yu PS (2017) Joint deep modeling of users and items using reviews for recommendation. In: Proceedings of the tenth ACM international conference on web search and data mining, pp 425–434
15. Chin JY, Zhao K, Joty S, Cong G (2018) ANR: Aspect-based neural recommender. In: Proceedings of the 27th ACM international conference on information and knowledge management, pp 147–156
16. Cheng Z, Ding Y, He X, Zhu L, Song X, Kankanhalli MS (2018) A ^{3}NCF: an adaptive aspect attention model for rating prediction. In: IJCAI, pp 3748–3754
17. Cheng Z, Ding Y, Zhu L, Kankanhalli M (2018) Aspect-aware latent factor model: rating prediction with ratings and reviews. In: Proceedings of the 2018 world wide web conference, pp 639–648
18. Liu B (2012) Sentiment analysis and opinion mining. Synthesis Lectures Hum Lang Technol 5(1):1–167
19. Pontiki M, Galanis D, Papageorgiou H, Androutsopoulos I, Manandhar S, Al-Smadi M, Al-Ayyoub M, Zhao Y, Qin B, De Clercq O et al (2016) Semeval-2016 task 5: aspect based sentiment analysis. In: International workshop on semantic evaluation, pp 19–30
20. Hazarika D, Poria S, Vij P, Krishnamurthy G, Cambria E, Zimmermann R (2018) Modeling inter-aspect dependencies for aspect-based sentiment analysis. In: Proceedings of the 2018 conference of the North American chapter of the Association for Computational Linguistics: human language technologies, Vol 2 (Short Papers), pp 266–270
21. Luo H, Ji L, Li T, Duan N, Jiang D (2020) Grace: gradient harmonized and cascaded labeling for aspect-based sentiment analysis. *arXiv preprint* arXiv:2009.10557
22. Ruder S, Ghaffari P, Breslin JG (2016) A hierarchical model of reviews for aspect-based sentiment analysis. *arXiv preprint* arXiv:1609.02745
23. Wang Y, Huang M, Zhu X, Zhao L (2016) Attention-based LSTM for aspect-level sentiment classification. In: Proceedings of the 2016 conference on empirical methods in natural language processing, pp 606–615
24. Lee-Thorp J, Ainslie J, Eckstein I, Ontanon S (2021) FNet: mixing tokens with fourier transforms. *arXiv preprint* arXiv:2105.03824
25. Liu P, Zhang L, Gulla JA (2021) Multilingual review-aware deep recommender system via aspect-based sentiment analysis. ACM Trans Inform Syst (TOIS) 39(2):1–33
26. Pennington J, Socher R, Manning CD (2014) Glove: global vectors for word representation. In: Proceedings of the 2014 conference on empirical methods in natural language processing (EMNLP), pp 1532–1543
27. Tai Y, Yang J, Liu X, Xu C (2017) MemNet: a persistent memory network for image restoration. In: Proceedings of the IEEE international conference on computer vision, pp 4539–4547
28. Liu Q, Zhang H, Zeng Y, Huang Z, Wu Z (2018) Content attention model for aspect based sentiment analysis. In: Proceedings of the 2018 world wide web conference, pp 1023–1032

Exploring Biomarker Identification and Mortality Prediction of COVID-19 Patients Using ML Algorithms

Rajan Singh and Prashant K. Srivastava

1 Introduction

The COVID-19 is a respiratory illness caused by a novel virus known as severe acute respiratory syndrome coronavirus 2 (SARS-CoV-2) that belongs to the *Coronaviridae* family of virus, named after the crown (Latin—*coronam*) like structure on the virus' outer surface. The World Health Organization (WHO) learned of the coronavirus on December 31, 2019, after reports of 'viral pneumonia' clusters emerged from Wuhan in China [1]. The disease was declared as a pandemic by the WHO on March 11, 2020. Subsequently, the virus has appeared in almost every country of the world and is still devastating some countries like India, Brazil, and U.S. with numerous re-emerging epidemic waves.

The most common symptoms of the COVID-19 are fever, dry cough, and fatigue. Other less common symptoms include loss of taste or smell, nasal congestion, conjunctivitis, sore throat, and headache. Some people that are identified as asymptomatic patients may not even show any symptoms at all, while others may become severely ill with shortness of breath, loss of appetite, confusion, persistent pain or pressure in the chest, and high body temperature [1–3]. The incubation time of the disease, i.e., the time between getting infected and the symptoms to appear is suggested to range from 2 to 10 days, while sometimes going to as high as 14 days. People aged above 60 or people with comorbidities like high blood pressure, heart–lung problems, diabetes, and cancer are more susceptible to developing serious illness [4, 5] and sometimes even death.

The disease transmits through human-to-human contact or human contact with the infected surfaces. The virus can also spread through the droplets resulting from

R. Singh (✉) · P. K. Srivastava
Department of Mathematics, Indian Institute of Technology Patna, Patna 801103, India
e-mail: rajan_1911mc10@iitp.ac.in; eternalrajan@gmail.com

P. K. Srivastava
e-mail: pksri@iitp.ac.in

coughing or sneezing of a COVID-19 infected person. The coronavirus' action mechanism most likely involves the use of the spike proteins present on the outer fatty layer to bind with a specific receptor called angiotensin-converting enzyme 2 (ACE2) present on the surface of the human cells. After fusing with the aforementioned receptor, the virus releases its genetic material, i.e., ribonucleic acid (RNA) into the human cell that causes the cell to produce copies of the virus, and thereby causing the cell death. Also, sometimes the immune cells of the body detect and destroy the infected cell. So, serious illness may occur, not because of the virus, but by the body's own immune system. Patients with the severe COVID-19 cases may experience the cytokine storm syndrome (CSS), a physiological reaction that causes the innate immune system to release excessive uncontrolled quantities of the cytokines. Cytokines are essential to fight foreign invaders; and work as an infection and inflammation signaling protein molecules in the human body. But, sometimes the cytokines can cause multi-organ failure or even death, in case of sudden excessive quantities. So, in order to reduce the chances of serious illness or fatality, the diagnosis and prognosis establishment of the disease for the infected patients should be done as early as possible.

The coronavirus is highly contagious and some reports even suggest that the virus may be airborne [6], thereby harder to control. With the passage of time, new variants of the coronavirus have appeared and a high possibility has arisen that more infectious and severe strains of the virus may emerge in the future. Thus, the highest priority of all the governments, the agencies and the people around the world is to apply various methods for ending the pandemic as soon as possible.

The ways to end the pandemic includes checking disease spread, disease detection/diagnosis and prognosis of the affected patients. The disease spread can be stopped by preventive measures like social-distancing mask sanitizers (SMS) and quarantine measures, while better patient care management can be carried out by rapid diagnosis and prognosis by evaluating the pertaining COVID-19 biomarkers. Thus, the study explores factors affecting the mortality of the COVID-19 patients using the available laboratory test measurements and other basic details. The problem was posed in the form of supervised ML classification task, with the laboratory test as the features and the mortality, i.e., final outcome, as the class label. The study identifies key factors, i.e., biomarkers of the COVID-19 disease, and discusses the developed mortality prediction model performance in context of the medical significance and healthcare settings. The early detection of the COVID-19 using the identified biomarkers can be helpful in easing the intense burden on the healthcare infrastructure.

2   Prior Research

The COVID ML related researches have been carried out with regards to the diagnosis and prognosis of the COVID-19 patients, since the time the disease was declared a pandemic [7, 8]. A ML-based research was carried out by Xie et al. [9], to propose a multivariate logistic regression (LR) classification model with internal validation for predicting the mortality of the COVID-19 patients. The healthcare data of 299 patients from the Tongji Hospital in Wuhan, China was used. Using backward stepwise feature selection (BSFS) and bootstrap resampling, a model was developed based on age, biomarkers and other comorbidities from the nine available features, with the discrimination assessment done using concordance-statistic. For calibration purposes, calibration-in-the-large, and calibration slopes and plots were used. Age, lactate dehydrogenase (LDH), lymphocyte, and peripheral oxygen saturation (SpO$_2$) were found to be independent predictors of mortality with a score of 0.89 for internal and 0.98 for external validation. The model performance was excellent considering the few selected features in predicting the mortality of the COVID-19 patients.

Wang et al. [10] had developed two separate predictive models for the in-hospital mortality of COVID-19 patients. The first model was clinical based, used age, heart disease, and hypertension history of the patient to give a AUCROC score of 0.83 on the validation set. The second model was laboratory based, used age, and other blood-related pathological markers like aspartate aminotransferase, D-dimer, high sensitivity C-reactive protein (hs-CRP), lymphocytes, neutrophil, glomerular filtration rate, and SpO$_2$ to give an AUCROC score of 0.88 on the validation set. The validation cohort consisted of 44 COVID-19 patients. For feature selection, eXtreme Gradient Boosting (XGBoost) and BSFS was used, and for classification multivariate LR was used.

Yan et al. [11] with the goal of producing an interpretable model for prediction of COVID-19 patients, used the data containing multiple blood sample readings of 485 patients from the aforementioned Tongji Hospital. The last available reading of the patients was used for the study, with carry from last observation missing value imputation technique. The researchers used XGBoost for mortality prediction and determined three important biomarkers—LDH, Lymphocytes and hs-CRP. Decision rules were made from the three selected features using the respective threshold values to determine the mortality of the patients. The model could predict the final outcome of the patient with a 90% accuracy for all days before the outcome and was able to discriminate between the mortality and the survivability of the patients.

Karthikeyan et al. [12] took the data from Yan et al. [11] study to form a ML based clinical decision support system for early prediction of COVID-19 mortality. First, K-nearest neighbors (KNN) imputation with K value set to ten was used to impute the missing data values of pre-processed 1766 multiple readings corresponding to 370 patients. Next, the splitting of data was done in the 80:20 ratio while maintaining the survived to dead ratio same for the train and test data. The XGBoost was used on the train data to get the important features and was fed into the neural network forward feature selection. Four features in addition to age were selected that included hs-CRP, LDH, LP, and NP. With the identified features, six different algorithms were trained and tested to obtain 96% accuracy with the two hidden layer neural network. The model was able to predict the outcome as early as 16 days before the outcome.

The discussed studies showed a good performance in determining the mortality prediction but does not properly consider the handling of the medical data used. For example, Karthikeyan et al. [12] directly applies the K-nearest neighbors (KNN) imputation technique without considering the medical significance of the features present. Also, temporal change effects on the patients' readings, that are a crucial part in healthcare scenarios, were not included during their data imputation step. For developing a model pertaining to the medical field, a proper medical data handling must be carried out first. So, in the carried-out study, a new method of data imputation was devised that was targeted to maintain the medical data quantity without affecting its quality. Further, the biomarkers related to the coronavirus disease were identified, and were later used for developing a mortality prediction model. The study's aim is akin with the referenced research but takes a rigorous account of the medical significance of the data rather than direct application of the ML algorithms. The direct application of ML on medical data is meaningless if no medical essence of the data and the results are taken into consideration, as in the medical field even a small decision can alter the patient's outcome. So, a significant part of the study involved the careful examination of each and every aspect of medical data that included studying of data features (pathological markers), grouping of data features at several levels, effect of temporal changes on the data features, and unique patient distribution based on the pathological markers and the demographics.

3 Analysis, Model and Predictions

3.1 Data Resource

The study is based on the data taken from the Yan et al. [11] research paper. Originally, the data consisted of the health record information of the COVID-19 patients that was collected between 10 January and 18 February 2020 from the Tongji Hospital in Wuhan, China. The data consisted of multiple laboratory test readings of 375 patients in addition to age, gender, hospital details, and the final outcome. Out of the 375 patients, 174 patients died, while 201 survived. Total number of columns were 81 including the outcome class label and the number of readings were 6120. Multiple blood sample readings were taken during the hospitalization of the patients. The readings were incomplete, i.e., not all tests were carried out for any particular sample reading. Since, the domain of the data pertained to the medical and health-care field, the data and the features were necessarily studied in the context of the medical terminology. All the features, i.e., laboratory tests were studied one by one, and the corresponding use and prior research were explored. The total number of features were 80, and were of three types, i.e., two demographic details, four hospital details, and 74 laboratory test measurements. The class label was the final outcome of the patient in binary form, as dead or survived. So, the total number of columns was 81. Multiple readings of the patients were taken during the course of the hospital stay; the data had complete demographic details. A very few of the hospital admission/discharge details were missing while the pathological tests details were present differently for different patients and different readings. Originally, the data

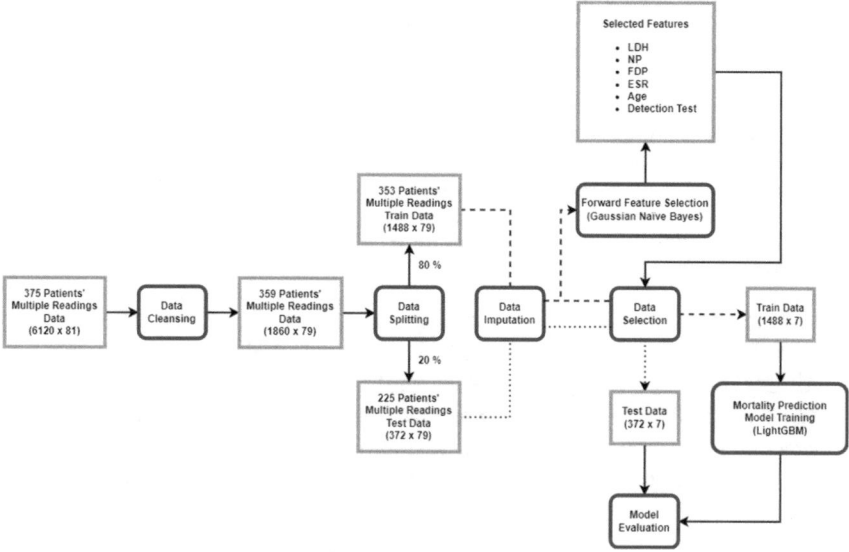

Fig. 1 Flowchart showing complete workflow of the model

had around 80% of the missing values and so, the data was not directly used, but after some pre-processing steps that included dropping of data points (further explained in Sects. 3.2 and 3.3). The complete workflow was carried out as shown in Fig. 1.

3.2 Data Cleaning

Of the 74 laboratory tests, four tests were related to other infectious diseases namely Hepatitis A, Hepatitis B, and human immunodeficiency virus acquired deficiency disease (HIV-AIDS), and syphilis. The effects of the four infectious diseases were not taken into account, as the data imputation technique could not be extended for the presence of the antibodies against the virus in non-infected patients. Thus, the four tests were dropped during the data cleaning step and so, the number of considered laboratory tests reduced from 74 to 70. The feature list had a detection test for the COVID-19 virus that was used to figure out whether the patients had become COVID-19 negative during the stay at the hospital, and also in determining the cause of patients' death, i.e., either due to COVID-19 as primary or due to the arising complications (secondary). Since, some recent studies suggest that once the patient becomes COVID-19 negative, the re-infection occurs only after at least 6 months [13]. So, the patients that became COVID-19 negative once, were assigned a negative value, i.e., -1, in the rest of the succeeding readings for the corresponding patients. Also, the samples' reading time were replaced with the available discharge time, if the latter were missing. After that, the following two additional temporal features were added for to be used in the later steps:

- Reading admission time difference (RATD)—Time difference between the sample reading and the admission time.
- Discharge reading time difference (DRTD)—Time difference between the discharge and the sample reading time.

Next, the features were divided into four types, on the basis of corresponding use in the further steps:

- Excluded features—four aforementioned infectious disease antibody tests, i.e., test for Hepatitis A, Hepatitis B, HIV-AIDS, and syphilis.
- Primary features—69 laboratory tests other than the excluded features.
- Primary additional features—Two added temporal features, one COVID-19 detection test, and two demographic features, i.e., age and gender.
- Additional features—four hospital details, i.e., admission time, discharge time, patient identification, and reading time.

First, the excluded features were dropped from the data, and further, the number of the readings with only positive DRTD and with at least 1 of the 69 primary features was considered for further steps. The number of readings was reduced to 5307 from 6120, and the number of unique patients dropped to 359 from 375. The total number of columns became 79. Except for the 69 primary features, all other remaining features were complete after the data cleansing.

3.3   Data Splitting

The data with 5307 samples was analyzed and the number of missing values with respect to the features were calculated, Fig. 2. The number of samples and unique patients was plotted with varying number of least features present, i.e., feature threshold. The analysis was done with only the 69 primary features, and showed that the increase in the feature threshold caused a significant decrease in missing data, samples, and unique patients' percentage. The threshold was set to 13, i.e., samples with at least 13 features out of the 69 primary features were selected, because of two reasons. First reason was that the value 13 lies near the middle of the plateau region of data present, samples and unique patients' percentage. Secondly, the maximum number of features in the manual categorization of the features during the data imputation step was 12, the feature threshold value was set to 13 so as to minimize the possibility of any sample to contain only that group's 12 features. Thus, the number of selected samples was 1860 that had around 33.96% non-null value data cells of the total 128340 present in the dataset. The number of unique patients considered was 356 out of the 359 present. Also, out of the 356 patients considered, 195 survived, 133 died due to COVID-19 as a primary cause while 28 died due to the complications caused by the virus.

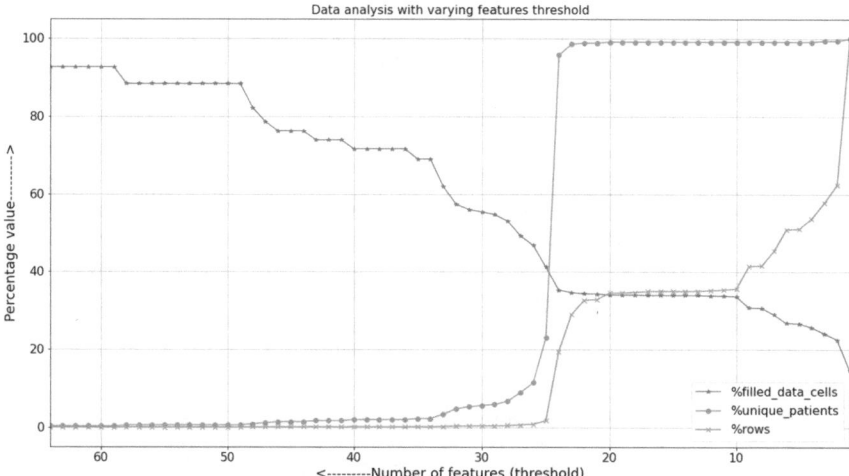

Fig. 2 Percentage of unique patients and data present with varying number of feature threshold

The data was split into train and test data in the 80:20 ratio in a stratified manner with respect to the outcome that ensured similar distribution of dead and survived patients in both train and test data. The train data had 1488 sample readings of 147 unique female and 206 unique male patients, i.e., 353 unique patients of 359 and the mean age of the patients was 58.73 ± 16.51 years. Out of the 353 patients in the trainset, 195 survived with 103 females and 92 males, while 44 females and 114 males were from 158 dead patients. The test data had 372 sample readings of 96 unique female and 129 unique male patients, i.e., 225 unique patients of 359. The mean age of the test data patients was 58.88 ± 16.25 years. Out of 225 patients, 128 survived with 69 females and 59 males, while 27 females and 70 males were from 97 dead patients.

3.4 Missing Data Handling

The missing data handling technique for medical data used, must ensure both the quality and quantity of the data. So, various missing data handling techniques were explored on the qualitative basis for the data used here. The following were the missing data handling techniques explored, while assessing the corresponding advantages and limitations [14] with respect to the data used Table 1.

After exploring the mentioned data handling techniques, a new imputation technique was developed and tried out on the data, Fig. 3. The technique was designed keeping in mind about all the limitations of the mentioned techniques. Since, the KNN imputation uses a similarity function to find the best value for imputation but suffers during high numbers of missing values, and with higher dimension data.

Table 1 Qualitative comparison of data handling

Technique	Mechanism	Advantages	Limitations
Dropping	Samples missing more than specified threshold number are dropped	• Simple	• Can lead to drastic reduction in the number of samples
Mean, median, mode or number replacement	Missing values are replaced with mean, median, mode, or some number	• Simple • No reduction in number of samples	• Neglects multiple patient readings • Neglect effects of temporal changes
Last observation carried forward or backward	Replace the missing value with previous or next known value	• No reduction in number of samples • Can consider multiple patient readings	• Neglect effects of temporal changes
KNN imputation	Replaces missing values by mean of similar neighboring sample values	• Can consider multiple patient readings • Can consider the effects of temporal changes	• Not suitable for data with a large number of missing values • Not suitable for high dimension data
Multiple imputation	Replaces each missing value by several values from a possible set and performs some analyses to find the best value	• Can consider the effects of temporal changes • Can work with data with a large number of missing values	• Not suitable for high dimension data • Not suitable for missing not at random data (MNAR)

Thus, the KNN imputation with a prior manual step was recursively carried out on the subset of features to eliminate those limitations. The following were the two steps for the conceived data imputation technique:

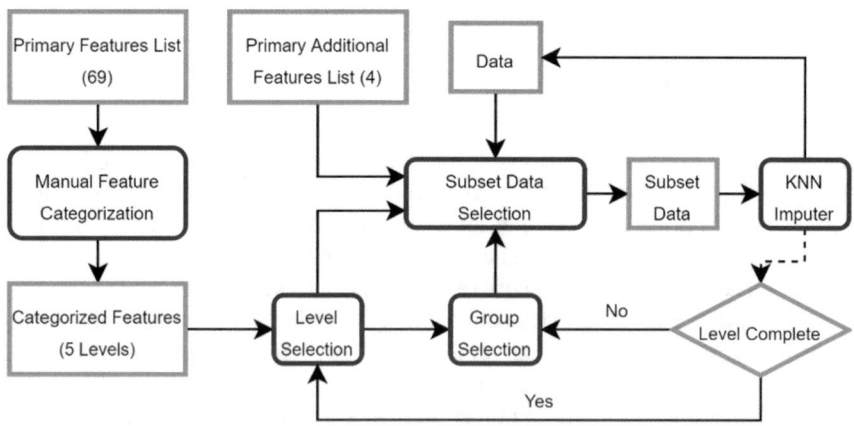

Fig. 3 Flowchart depicting data imputation technique developed

- Manual feature categorization
- Recursive subset KNN data imputation

Manual Feature Categorization: The first step for the missing data imputation involved a manual categorization of the features into groups at several levels on the basis of some similarity. The features, i.e., laboratory tests, are generally used for determining some specific diseases or dysfunctions. Each test involves some kind of biomolecule or cellular measurement. Several types of biomolecules and cells are constantly produced by the human body for proper functioning and maintaining healthy equilibrium. Thus, the similarity factors for feature grouping were based on the corresponding biomolecule or cellular measurement used in the tests. Following were the three similarity factors to determine the groups:

- Biomolecule/Cell Function: Each biomolecule and cell perform some kind of function like fighting infection, maintaining metabolic processes, nutrient transportation, etc., and so slightest change in the quantity affects the equilibrium of the healthy body. Thus, the specific response and functions were considered as the most significant factor of all three. The function factor could be generic or specific.
- Biomolecule/Cell Type: Several types of biomolecules like carbohydrates, lipids, proteins, etc., and the cells like erythrocytes, leukocytes, thrombocytes, etc., are produced inside the human body. So, the type of the biomolecule or the cell also plays an important role for grouping criteria. The type factor could be of biomolecule, cellular, or any other form.
- Biomolecule/Cell Site: The origin source and the action location of the biomolecule or the cell generally can be used for the identification of any problem within or nearby the source itself. Thus, the source and action site were also included in the factors. The site factor could be broad or specific.

So, these factors were used to determine the placement of a feature in the groups. In case of multi-function factor, the type and the corresponding origin source were used in combination. Levels were created on the basis of these three factors, and so, the grouping also varied with regards to the level. Following were the levels that were created:

- Level 0/Base Level: In the base level, all features were placed in one group without considering any similarity factors.
- Level 1/Organ System Level: The organ system level was constructed on the basis of the site factor in broad perspective. The human body has 11 major organ systems, namely, circulatory system, digestive and excretory system, endocrine system, integumentary system, lymphatic system, muscular system, nervous system, renal system, reproductive system, respiratory system, and the skeletal system. The organ systems are not mutually exclusive and are interconnected with each other.
- Level 2/Organ Level: The organ systems involve the working of individual interconnected organs. Thus, the organ level was set up on the basis of a specific

organ with the pertaining function. The site factor in the specific perspective of the feature was also used for the grouping purpose.

- Level 3/Function Level: The function level involves the use of two factors, namely, the function factor and the site factor for determining the corresponding group of the feature. The two factors used considered both generic and specific perspectives.
- Level 4/Response Level: The response level was constructed using all three similarity factors and the prior levels' information. The feature attributes for all three factors included only the specific values and not the generic ones.

Recursive Subset KNN Data Imputation: The KNN algorithm is a supervised ML technique used for both classification and regression purposes. The KNN computes the distance information from a given number of neighboring data, i.e., k value to predict the label or the value. For continuous data, the algorithm uses mean, and for discrete data the algorithm uses most frequent value, from the given number of nearest neighbors. The algorithm is a non-parametric algorithm, i.e., devoid of any assumptions about the data. Some selected form of distance metric like cosine distance, euclidean distance, manhattan distance, etc., for continuous data and hamming distance for discrete data is used by the algorithm for value prediction.

In similar ways, the KNN data imputation works by using the data from the given number of neighbors to calculate the missing values. The KNN imputer with the distance weighted euclidean metric, i.e., weights are assigned as inverse of their distance that causes close neighbors to influence more, was used for data imputation. The imputation starts at the highest level, i.e., response level or level 4. For each group of the level, a subset was created using only the samples with at least one feature from the 69 primary features and one features from the primary additional features excluding RATD. Further, features that had all missing values were dropped from the subset that were filled in other later iterations. Normalization of the subset data using the min–max technique was carried out. Next, the KNN imputation was carried out for all the groups with the number of neighbors k set to the cube root of the number of samples in that subset.

Once all the data values of a group subset were imputed, the data was renormalized again to retain the original known values that were to be used later for other level iterations. After level 4 completion, data imputation was carried out for subsequent levels from top to bottom in the same recursive subset manner. All 84,758 missing values of the total of 128,340 values were filled using the mentioned technique.

3.5 Feature Selection

Biomarkers are some forms of measurable indicators used for disease diagnosis, patient prognosis, and treatment effectiveness. Biomarker indicators may be biological, chemical, or physical, and the corresponding measurement may be biochemical, cellular, functional, physiological, or molecular. Thus, sequential feature selection (SFS), i.e., a wrapper method for dimensionality reduction in ML problems, was used to identify the biomarkers for the COVID-19 disease from the 69 primary features from only the train data with 1488 sample readings.

Sequential Forward Feature Selection: SFS are a type of greedy search algorithm used for feature reduction by selecting a subset of relevant features from the set of considered features. SFS reduces the need of expensive computational requirements and removes any irrelevant feature if present. From the available types of SFS, sequential forward feature selection (SFFS) was used. Initially, SFFS starts with an empty feature set and afterwards greedy selects a feature in every iteration while retaining the previously best selected feature subset. After that, the selected new feature is added to the subset and the iterations are carried until the specified number of features are selected. The algorithm uses an estimator to fit the data and measure the performance during the iterations to select the best feature in combination with the previous selected feature subset.

SFFS Estimator: The estimator used in SFFS was Gaussian naïve Bayes (GNB) classifier with stratified five-fold cross-validation for performance evaluation in each iteration. GNB is an extension of naïve Bayes algorithm for the continuous data. The naïve Bayes algorithm uses a probabilistic approach based on the Bayes' theorem that represents the probability of an event using prior knowledge of the conditions related to that event. The naïve Bayes algorithm naïvely assumes that the features are strongly independent from each other, hence the name. The GNB classifier assumes that the data follows a Gaussian distribution. The SFFS estimator was used over various other estimators because of the following reasons:

- The variation in the performance for the GNB was significantly low as compared to other estimators.
- GNB has almost no hyperparameter unlike other estimators, so the feature selection process was independent of the hyperparameter values.
- GNB was way faster than the other estimators during the feature selection.

During the SFFS with GNB as the estimator, instead of starting with an empty feature subset, age and the COVID-19 detection test were included in the first subset. Age was a fixed feature to compensate for the effect of comorbidities, as older people are more prone to developing serious illnesses [5] and having other comorbidities. The COVID-19 detection test was also fixed so as to gage the effect of people dying because of the virus or due to the virus' complications. Also, gender was included to assess the effect played in the role of the COVID-19 disease, as some diseases are known to have some gender dependence. So, the total number of considered features was 73 with 69 primary features, age and COVID-19 detection test as fixed feature, and one gender feature. The SFFS was carried out for the complete range of subset features from 3 to all 73 features, i.e., two fixed features, gender and the 69 primary features. The performance of the estimator was plotted against the number of features in the subset, Fig. 4.

The performance accuracy of the estimator during the feature selection peaked with around 95% and saturated at a number of features equal to 9 and decreased after 16. In healthcare settings, fast and accurate tests are needed to increase the

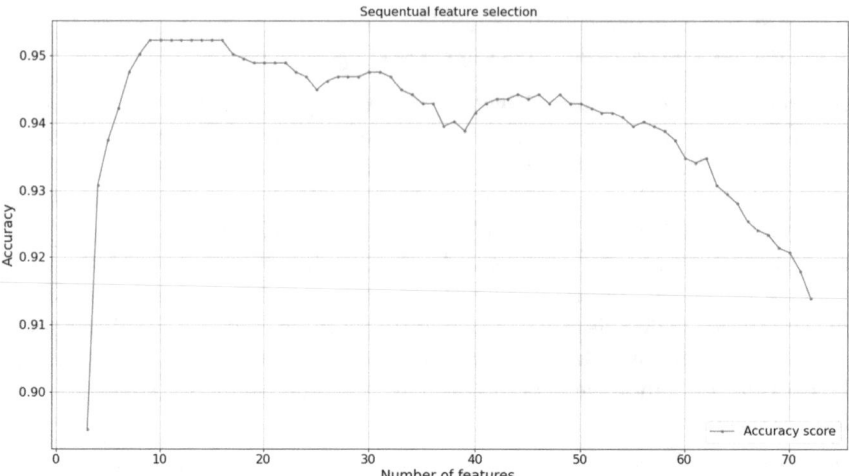

Fig. 4 Estimator's performance with varying number of selected features

survivability of the patients as time is a crucial factor. So, only the first top six features during the SFFS were considered for further steps. The six features showed an accuracy above 94% that was in between the lowest accuracy of 89% with three features and 95% accuracy with nine features. Excluding the two features of age and the COVID-19 detection test, following were the four features selected in order:

- Lactate dehydrogenase
- Neutrophils percentage
- Fibrin degradation products
- Erythrocyte sedimentation rate.

4 Model Development

The selected six aforementioned features were used for the mortality prediction of the COVID-19 patients. To develop a mortality prediction, an ensemble learning (EL)-based algorithm was selected. The gradient boosted trees (GBT) are one of the many EL-based algorithms that generally outperforms decision trees, random forests (RF), support vector machines (SVM), and several other algorithms. GBT uses an ensemble of weak DT to build a strong learning model that sequentially tries to improve upon the performance of the previous DT step by step. Light gradient boosting machine (LightGBM) is a highly efficient GBT framework developed by the Microsoft Corporation [15]. LightGBM uses two techniques, namely, gradient-boosted one-sided sampling and the exhaustive feature bundling. The algorithm splits the DT leaf-wise instead of level-wise splitting as used in other boosting algorithms, thereby sometimes even outperforming the state-of-the-art XGBoost algorithm. LightGBM grows the tree leaf that shows maximum decrease in the loss function; and uses a highly optimized histogram-based DT algorithm that causes less time complexity. Thus, LightGBM was used with the selected six features to build a mortality prediction model for the COVID-19 patients. Also, to show the comparative performance, five other algorithms were compared with the LightGBM based model using the selected six features, Fig. 5. The six algorithms in the increasing order of the performance are shown in Table 2.

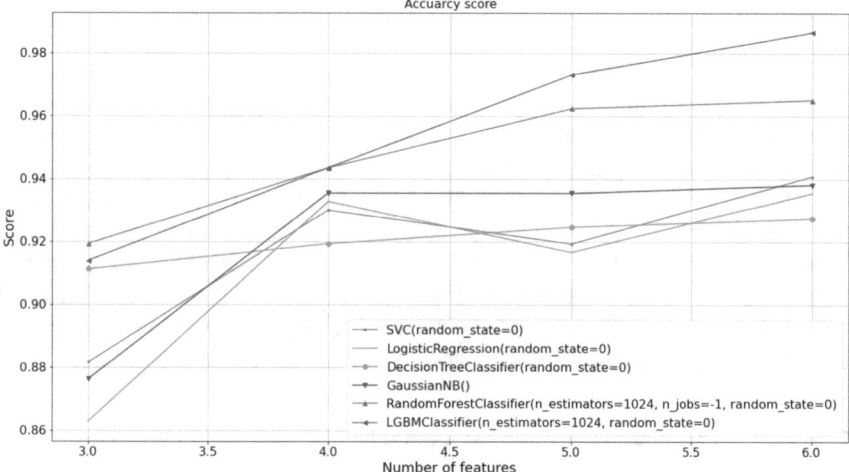

Fig. 5 Algorithms' performance with the variation in number of selected top features

Table 2 Performance comparison of algorithms with selected six features

Name	Accuracy (%)	AUCROC (%)
Decision trees	92.74	93
Logistic regression (normalized data)	93.55	98
Gaussian Naïve Bayes	93.82	98
Support vector machines (normalized data)	94.09	98
Random forest	96.50	99
Light gradient boosting machine	98.66	100

5   Results and Discussions

5.1   Interpretation of Identified Biomarkers

The identified COVID-19 biomarkers using the feature selection techniques were analyzed and interpreted on the basis of medical significance. The selected six features included four laboratory tests, age, and the virus detection test. Further analysis with respect to time was carried out for all the four laboratory tests and the corresponding medical significance gave useful insight about the coronavirus and the COVID-19 disease.

Lactate Dehydrogenase: The first feature selected during the feature selection was lactate dehydrogenase in combination with age and detection test. LDH is an enzyme that is found in all living cells and is released during tissue damage and inflammation inside the body. The distribution plot of the LDH with respect to the days for the final outcome suggests that the surviving patients had significantly lower values of LDH throughout hospital stay, as compared to the patients that died either from COVID-19 or from the arising complications. Prior studies [9, 11, 12] on the LDH and the coronavirus also suggest that since the coronavirus hijacks the cell and causes cell damage, high quantities of LDH are released, explaining the elevated levels, Figs. 6 and 10 (left).

Neutrophils Percentage: Neutrophils are a type of leukocytes and help in fighting off viruses, bacteria, and other foreign invaders by the process of chemotaxis, degranulation, and phagocytosis. Neutrophils have been reported to be a biomarker associated with the progression of COVID-19 patients [16]. Also, other studies points, that

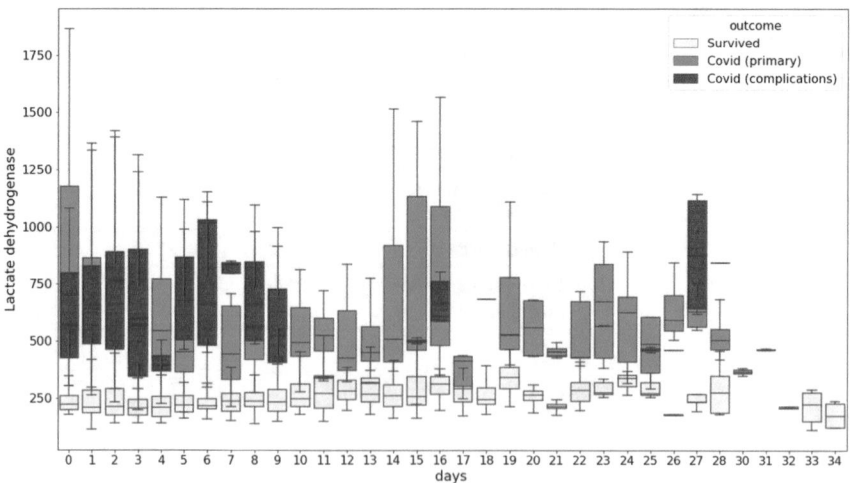

Fig. 6 LDH variation with time to final outcome

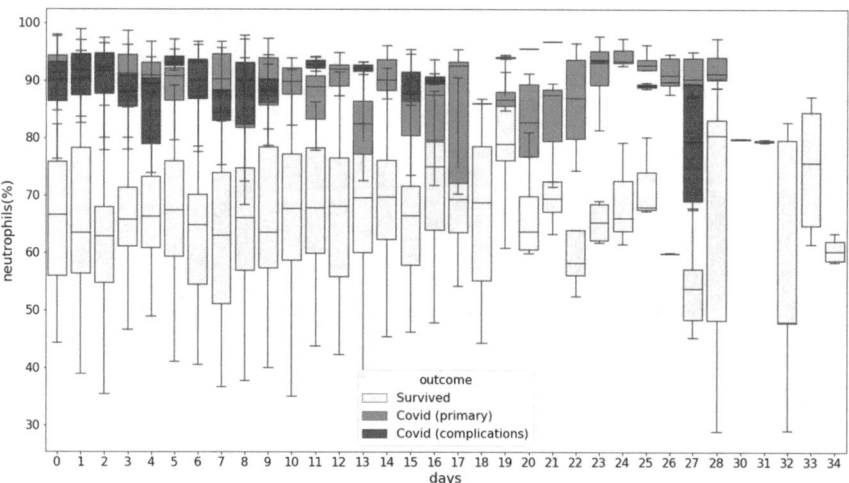

Fig. 7 NP variation with time to final outcome

neutrophils in combination with LDH could be useful in the determination of COVID-19 severity [17]. Neutrophils also showed a similar trend in terms of the distribution plot as with the LDH, with the surviving patients having lower neutrophils percentage as compared to the dead patients, Figs. 7 and 10 (right).

Fibrin Degradation Products: Fibrin degradation products are produced during the blood clot degradation and the corresponding test is usually conducted for diagnosis of disseminated intravascular coagulation (DIC). DIC refers to the occurrence of blood clots throughout the body that blocks small blood vessels. The prevalence of blood clots has been found not only in COVID-19 patients and even in the COVID-19 negative recovering patients. Some studies have shown that the FDP may affect the disease severity [18]. The distribution plot with respect to time didn't show a clear demarcation for survived and dead patients, but the survived patients had lower values of FDP as compared to the dead patients overall, Figs. 8 and 11 (left).

Erythrocyte Sedimentation Rate: The last feature to be selected was the erythrocyte sedimentation rate (ESR). ESR test provides an identification of inflammation present in the body but does not specify the condition; so, other tests are carried with ESR to determine the exact reason. Although a clear differentiation was not present for dead and survived patients, the patients that died due to the COVID-19 as a primary cause had lower values of the ESR near the days to final outcome. Some observations have been made in a case report [19] about elevated ESR in patients, even after recovering from the disease, Figs. 9 and 11 (right). Since the ESR was selected last, a combination with other previous selected tests might help in better early prognosis of the COVID-19 patients.

The demarcation of the dead and surviving patients obtained from the individual distribution plot became less prominent, as more and more features were selected.

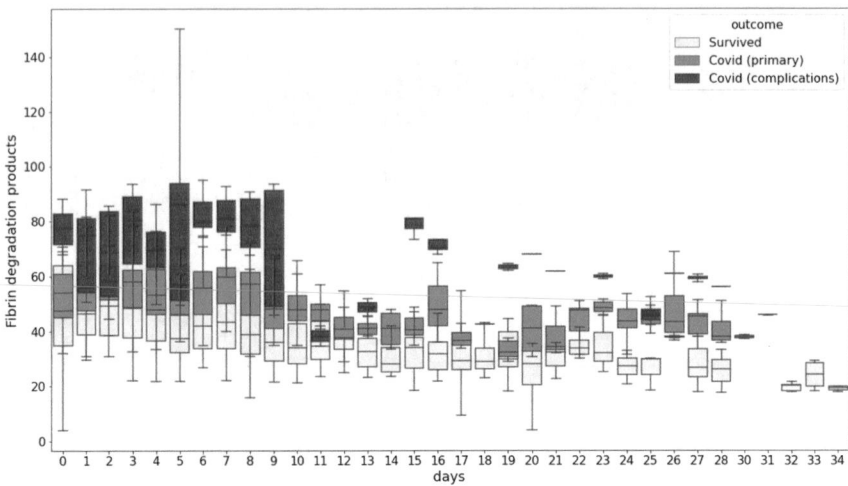

Fig. 8 FDP variation with time to final outcome

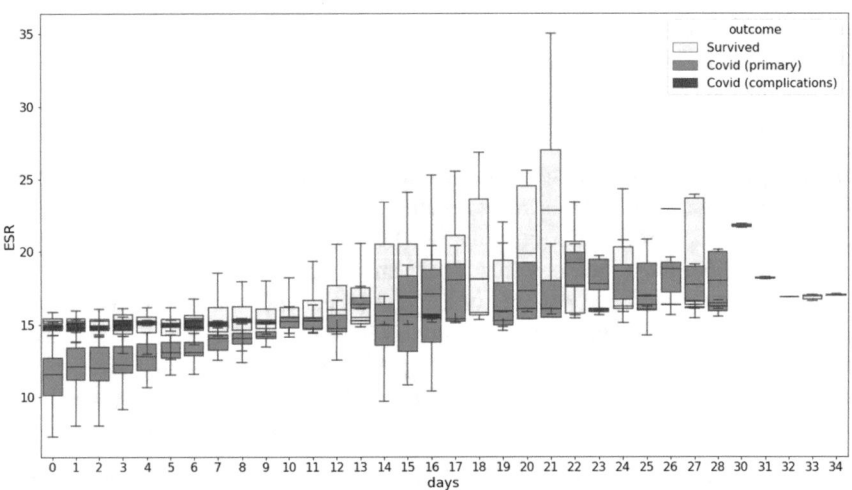

Fig. 9 ESR variation with time to final outcome

So, the latter selected features may be better utilized in combination with the former selected features for better disease prognosis and mortality prediction.

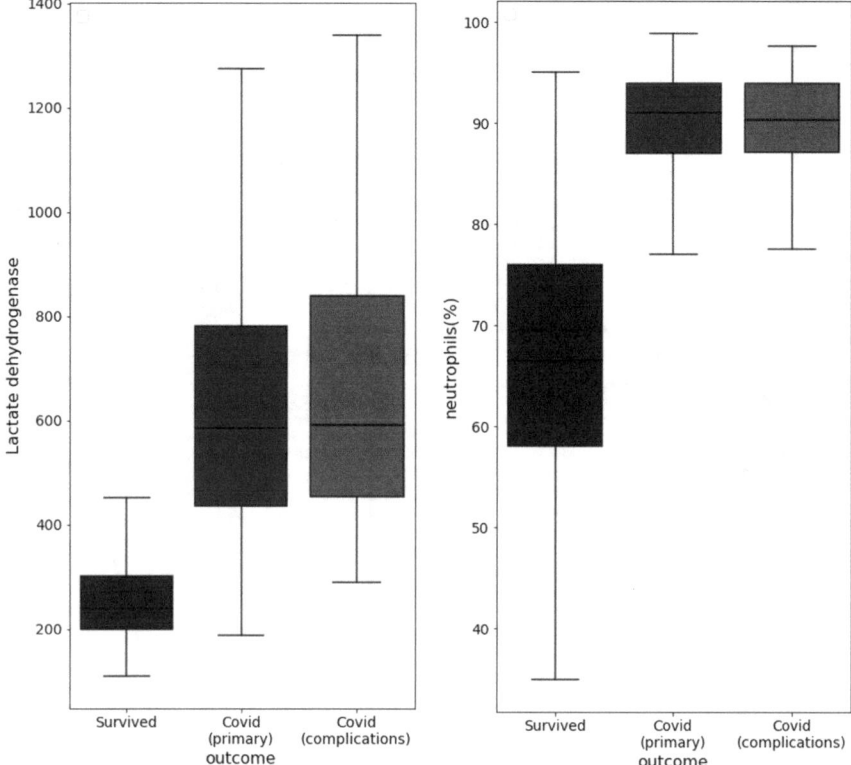

Fig. 10 (Left) LDH overall distribution and (right) NP overall distribution

5.2 Performance Analysis of Mortality Prediction Model

The model developed for the mortality prediction using the LightGBM, performed
extremely well on the test data while outperforming the other five algorithms. The
model was able to identify the patients that were more likely to die from the disease
using only the selected six features. The accuracy obtained was 98.66% with 98.63%
F1 Score, 97.83% precision, 99.45% sensitivity, and 97.91% specificity values on
the test data, Fig. 12. The temporal analysis of the model performance showed that
the model was able to correctly predict the final outcome on an average of 8 days in
advance with more correct prediction skewed toward the near final outcome, Fig. 13.
The skewness of the temporal plot may be due the large number of readings taken
during the days near to the final outcome. For surviving patients, the model was able
to correctly predict the outcome on an average of 9 days in advance; and for dead
patients, on an average of 6 days in advance. Thus, the mortality predictive ability
of the model for the COVID-19 patients was excellent.

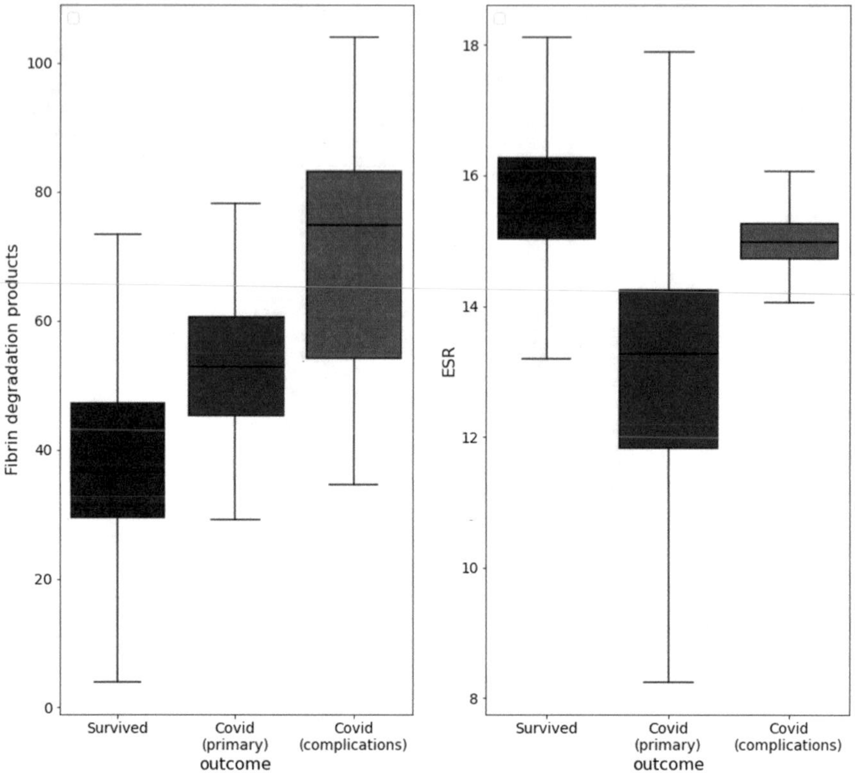

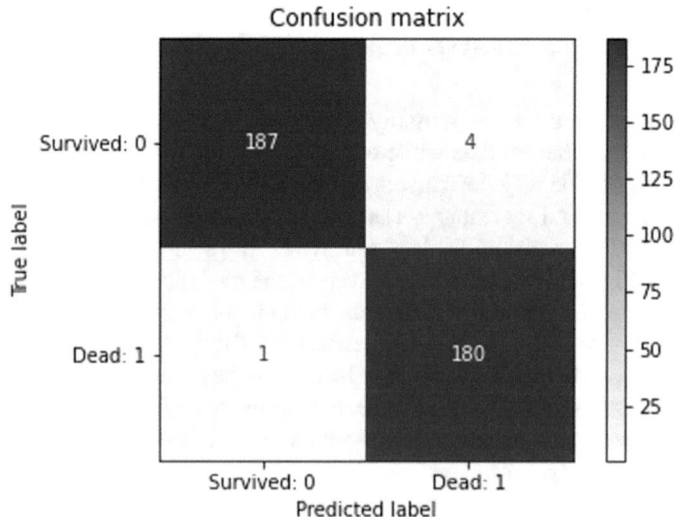

Fig. 11 (Left) FDP overall distribution and (right) ESR overall distribution

Fig. 12 Confusion matrix from model testing on 225 unique patients (multiple readings)

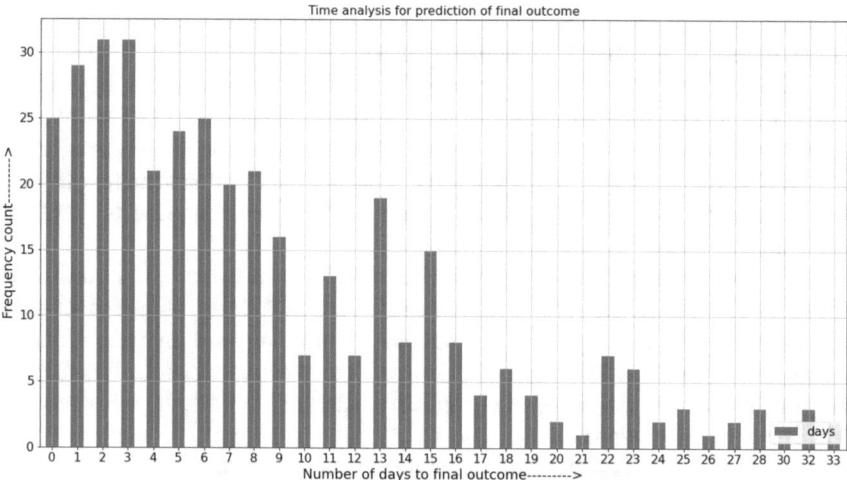

Fig. 13 Correctly-predicted frequency count with number of days to final outcome

5.3 Limitations

The developed model showed great performance in determining the mortality of the patients, but had some certain limitations. First limitation occurred with the data; as the data had a lot of missing values, the model was developed and tried only on the simulated data. Secondly, the effect of comorbidities and other infectious diseases was not considered due to the unavailability of the proper related data, even though age was used as a fixed feature, in order to compensate for the comorbidities. The absence of treatment policies followed for the patients, also posed some form of limitation to the work. So, external data is required for getting the complete picture out of the carried-out work. Also, the appearance of new variants [20] of the virus may show an unknown effect on the performance of the model depending on whether the mutated virus shows any significantly different pathological characteristics.

6 Conclusion and Future Scope

The identification of COVID-19 biomarkers can be quite useful in reducing the immense burden on the healthcare systems. The handful strong combination of six selected features can be utilized in the early and accurate prognosis of the COVID-19 patients. The four selected tests, namely, lactate dehydrogenase, neutrophils, fibrin degradation products, and erythrocyte sedimentation rate, and the corresponding medical interpretation can help not only in the disease prognosis but also in therapeutic management based on the effect of the drugs used in the treatment of the COVID-19 patients. The application of the developed mortality prediction model in healthcare settings can enable the healthcare workers to save precious human lives by making quick and effective treatment policies.

And since AI is revolutionizing the whole world, the possibility that 'the techno-logical singularity' event may occur in the near future, is very high. The use of AI in the healthcare field has seen rapid growth, and the recent pandemic has accelerated that growth. So, the study was carried out with the intent to contribute a little in the further expansion of AI in the healthcare field and in countering the ongoing pandemic.

The future scope of the work may involve the effects of comorbidities and other infectious diseases on the severity and fatality of the COVID-19. More work on the imputation method developed may be carried out in future work by refinements in classification of feature levels/grouping, in-depth understanding of the functions of the pathological markers used, and quantitative comparison with the other medical data imputation techniques. Also, the effect of the treatment, i.e., medicines and therapies given to the patients may be included in the future work.

Acknowledgements Authors are thankful to anonymous referees for their valuable suggestions to improve the manuscript. The research by Rajan Singh was financially supported by the institute's (Indian Institute of Technology Patna) scholarship.

References

1. World Health Organization: Coronavirus Disease 2019 (COVID-19) Situation Report—1 (2020). https://www.who.int/docs/default-source/coronaviruse/situation-reports/20200121-sitrep-1-2019-ncov.pdf
2. World Health Organization: Coronavirus Disease 2019 (COVID-19) Situation Report—51 (2020). https://www.who.int/docs/default-source/coronaviruse/situation-reports/20200311-sitrep-51-covid-19.pdf
3. Liang W, Liang H, Ou L, Chen B, Chen A, Li C, Li Y, Guan W, Sang L, Lu J, Xu Y, Chen G, Guo H, Guo J, Chen Z, Zhao Y, Li S, Zhang N, Zhong N, He J (2020) Development and validation of a clinical risk score to predict the occurrence of critical illness in hospitalized patients with COVID-19. JAMA Intern Med 180:1081. https://doi.org/10.1001/jamainternmed.2020.2033
4. Bajgain K, Badal S, Bajgain B, Santana M (2021) Prevalence of comorbidities among individuals with COVID-19: a rapid review of current literature. Am J Infect Control 49:238–246. https://doi.org/10.1016/j.ajic.2020.06.213
5. Liu Y, Mao B, Liang S, Yang J, Lu H, Chai Y, Wang L, Zhang L, Li Q, Zhao L, He Y, Gu X, Ji X, Li L, Jie Z, Li Q, Li X, Lu H, Zhang W, Song Y, Qu J, Xu J (2020) Association between age and clinical characteristics and outcomes of COVID-19. Eur Respir J 55:2001112. https://doi.org/10.1183/13993003.01112-2020
6. Greenhalgh T, Jimenez J, Prather K, Tufekci Z, Fisman D, Schooley R (2021) Ten scientific reasons in support of airborne transmission of SARS-CoV-2. Lancet 397:1603–1605. https://doi.org/10.1016/S0140-6736(21)00869-2
7. Lalmuanawma S, Hussain J, Chhakchhuak L (2020) Applications of machine learning and artificial intelligence for Covid-19 (SARS-CoV-2) pandemic: a review. Chaos Solitons Fractals 139:110059. https://doi.org/10.1016/j.chaos.2020.110059
8. Rampton V, Mittelman M, Goldhahn J (2020) Implications of artificial intelligence for medical education. Lancet Digit Health 2:e111–e112. https://doi.org/10.1016/S2589-7500(20)30023-6

9. Chow D, Glavis-Bloom J, Soun J, Weinberg B, Loveless T, Xie X, Mutasa S, Monuki E, Park J, Bota D, Wu J, Thompson L, Boden-Albala B, Khan S, Amin A, Chang P (2020) Development and external validation of a prognostic tool for COVID-19 critical disease. PLoS ONE 15:e0242953. https://doi.org/10.2139/ssrn.3562456

10. Wang K, Zuo P, Liu Y, Zhang M, Zhao X, Xie S, Zhang H, Chen X, Liu C (2020) Clinical and laboratory predictors of in-hospital mortality in patients with coronavirus disease-2019: a cohort study in Wuhan, China. Clin Infect Dis 71:2079–2088. https://doi.org/10.1093/cid/cia a538

11. Yan L, Zhang H, Goncalves J, Xiao Y, Wang M, Guo Y, Sun C, Tang X, Jing L, Zhang M, Huang X, Xiao Y, Cao H, Chen Y, Ren T, Wang F, Xiao Y, Huang S, Tan X, Huang N, Jiao B, Cheng C, Zhang Y, Luo A, Mombaerts L, Jin J, Cao Z, Li S, Xu H, Yuan Y (2020) An interpretable mortality prediction model for COVID-19 patients. Nat Mach Intell 2:283–288. https://doi.org/10.1038/s42256-020-0180-7

12. Karthikeyan A, Garg A, Vinod P, Priyakumar U (2021) Machine learning based clinical decision support system for early COVID-19 mortality prediction. Front Public Health 9. https://doi.org/10.3389/fpubh.2021.626697

13. Hall V, Foulkes S, Charlett A, Atti A, Monk E, Simmons R, Wellington E, Cole M, Saei A, Oguti B, Munro K, Wallace S, Kirwan P, Shrotri M, Vusirikala A, Rokadiya S, Kall M, Zambon M, Ramsay M, Brooks, T, BrownC, Chand, M, Hopkins, S, Andrews, N, Atti, A, Aziz, H, Brooks, T, Brown C, Camero D, Carr C, Chand M, Charlett A, Crawford H, Cole M, Conneely J, D'Arcangelo S, Ellis J, Evans S, Foulkes S, Gillson N, Gopal R, Hall L, Hall V, Harrington P, Hopkins S, Hewson J, Hoschler K, Ironmonger D, Islam J, Kall M, Karagiannis I, Kay O, Khawam J, King E, Kirwan P, Kyffin R, Lackenby A, Lattimore M, Linley E, Lopez-Bernal J, Mabey L, McGregor R, Miah S, Monk E, Munro K, Naheed Z, Nissr A, O'Connell A, Oguti B, Okafor H, Organ S, Osbourne J, Otter A, Patel M, Platt S, Pople D, Potts K, Ramsay M, Robotham J, Rokadiya S, Rowe C, Saei A, Sebbage G, Semper A, Shrotri M, Simmons R, Soriano A, Staves P, Taylor S, Taylor A, Tengbe A, Tonge S, Vusirikala A, Wallace S, Wellington E, Zambon M, Corrigan D, Sartaj M, Cromey L, Campbell S, Braithwaite K, Price L, Haahr L, Stewart S, Lacey E, Partridge L, Stevens G, Ellis Y, Hodgson H, Norman C, Larru B, Mcwilliam S, Winchester S, Cieciwa P, Pai A, Loughrey C, Watt A, Adair F, Hawkins A, Grant A, Temple-Purcell R, Howard J, Slawson N, Subudhi C, Davies S, Bexley A, Penn R, Wong N, Boyd G, Rajgopal A, Arenas-Pinto A, Matthews R, Whileman A, Laugharne R, Ledger J, Barnes T, Jones C, Botes D, Chitalia N, Akhtar S, Harrison G, Horne S, Walker N, Agwuh K, Maxwell V, Graves J, Williams S, O'Kelly A, Ridley P, Cowley A, Johnstone H, Swift P, Democratis J, Meda M, Callens C, Beazer S, Hams S, Irvine V, Chandrasekaran B, Forsyth C, Radmore J, Thomas C, Brown K, Roberts S, Burns P, Gajee K, Byrne T, Sanderson F, Knight S, Macnaughton E, Burton B, Smith H, Chaudhuri R, Hollinshead K, Shorten R, Swan A, Shorten R, Favager C, Murira J, Baillon S, Hamer S, Gantert K, Russell J, Brennan D, Dave A, Chawla A, Westell F, Adeboyeku D, Papineni P, Pegg C, Williams M, Ahmad S, Ingram S, Gabriel C, Pagget K, Cieciwa P, Maloney G, Ashcroft J, Del Rosario I, Crosby-Nwaobi R, Reeks C, Fowler S, Prentice L, Spears M, McKerron G, McLelland-Brooks K, Anderson J, Donaldson S, Templeton K, Coke L, Elumogo N, Elliott J, Padgett D, Mirfenderesky M, Cross A, Price J, Joyce S, Sinanovic I, Howard M, Lewis T, Cowling P, Potoczna D, Brand S, Sheridan L, Wadams B, Lloyd A, Mouland J, Giles J, Pottinger G, Coles H, Joseph M, Lee M, Orr S, Chenoweth H, Auckland C, Lear R, Mahungu T, Rodger A, Penny-Thomas K, Pai S, Zamikula J, Smith E, Stone S, Boldock E, Howcroft D, Thompson C, Aga M, Domingos P, Gormley S, Kerrison C, Marsh L, Tazzyman S, Allsop L, Ambalkar S, Beekes M, Jose S, Tomlinson J, Jones A, Price C, Pepperell J, Schultz M, Day J, Boulos A, Defever E, McCracken D, Brown K, Gray K, Houston A, Planche T, Pritchard Jones R, Wycherley D, Bennett S, Marrs J, Nimako K, Stewart B, Kalakonda N, Khanduri S, Ashby A, Holden M, Mahabir N, Harwood J, Payne B, Court K, Staines N, Longfellow R, Green M, Hughes L, Halkes M, Mercer P, Roebuck A, Wilson-Davies E, Gallego L, Lazarus R, Aldridge N, Berry L, Game F, Reynolds T, Holmes C, Wiselka M, Higham A, Booth M, Duff C, Alderton J, Jory H, Virgilio E, Chin T, Qazzafi M, Moody A, Tilley R, Donaghy T, Shipman K, Sierra R, Jones N, Mills G, Harvey D, Huang

Y, Birch J, Robinson L, Board S, Broadley A, Laven C, Todd N, Eyre D, Jeffery K, Dunachie S, Duncan C, Klenerman P, Turtle L, De Silva T, Baxendale H, Heeney J (2021) SARS-CoV-2 infection rates of antibody-positive compared with antibody-negative health-care workers in England: a large, multicentre, prospective cohort study (SIREN). Lancet 397:1459–1469. https://doi.org/10.1016/S0140-6736(21)00675-9

14. Schmitt P, Mandel J, Guedj M (2015) A comparison of six methods for missing data imputation. J Biomet Biostat 6:224. https://doi.org/10.472/2155-6180.1000224. https://www.hil arispublisher.com/open-access/a-comparison-of-six-methods-for-missing-data-imputation-2155-6180-1000224.pdf

15. LightGBM: a highly efficient gradient boosting decision tree (2017). In: Proceedings of the 31st international conference on neural information processing systems. Curran Associates Inc. https://doi.org/10.5555/3294996.3295074

16. Ponti G, Maccaferri M, Ruini C, Tomasi A, Ozben T (2020) Biomarkers associated with COVID-19 disease progression. Crit Rev Clin Lab Sci 57:389–399. https://doi.org/10.1080/10408363.2020.1770685

17. Fan X, Zhu B, Nouri-Vaskeh M, Jiang C, Feng X, Poulsen K, Baradaran B, Fang J, Ade E, Sharifi A, Zhao Z, Han Q, Zhang Y, Zhang L, Liu Z (2021) Scores based on neutrophil percentage and lactate dehydrogenase with or without oxygen saturation predict hospital mortality risk in severe COVID-19 patients. Virol J 18. https://doi.org/10.1186/s12985-021-01538-8

18. Asakura H, Ogawa H (2020) Overcoming bleeding events related to extracorporeal membrane oxygenation in COVID-19. Lancet Respir Med 8:e87–e88. https://doi.org/10.1016/S2213-2600(20)30467-7

19. Pu S, Zhang X, Liu D, Ye B, Li J (2021) Unexplained elevation of erythrocyte sedimentation rate in a patient recovering from COVID-19: a case report. World J Clin Cases 9:1394–1401. https://doi.org/10.12998/wjcc.v9.i6.1394

20. World Health Organization: Coronavirus Disease 2019 (COVID-19) Situation Report—50 (2020). https://cdn.who.int/media/docs/default-source/searo/indonesia/covid19/external-sit uation-report-50_7-april-2021.pdf?sfvrsn=888ff8eb_5

COVID-19 Cases Prediction Based on LSTM and SIR Model Using Social Media

Aakansha Gupta and Rahul Katarya

1 Introduction

The novel coronavirus epidemic originated in Wuhan, China, has rapidly spread in almost every country worldwide, and has produced a significant public health disaster across the world [1]. Due to the lack of understanding about the disease's progress in the early stages, the COVID-19 was not successfully suppressed. COVID-19 disease was later discovered to be infectious. The COVID-19 outbreak exemplifies the essential influence of this online information environment. As the disease spreads, there is a greater public awareness about COVID-19 on internet platforms such as Twitter, Instagram, Reddit, and so on [2–4]. The transmission of information can significantly impact people's behavior and the efficacy of government countermeasures. Online news was also a popular medium for disseminating COVID-19 updates and prevention and control information. Despite the public awareness and prevention of the coronavirus pandemic, the epidemic situation in the country was terrible. As a result, a study on the emergence and spread of epidemics is required.

With the emergence of a large-scale outbreak of infectious disease and declaration of a significant public health emergency, the public uses epidemiological models to estimate and anticipate the growth trend of the disease, and the results are used to drive the formation of preventative and control measures [5–8]. The most common classic epidemiological models are SI, SIR, and SEIR, where 'E' refers to the exposed population [9]. The government's preventative and control actions, as well as public knowledge, all had a part in the spread of the pandemic metal. The availability of transparent pandemic reporting and preventive and control measures has hastened the virus's spread [10]. As a result, pandemic data alone are not enough to provide reliable

A. Gupta (✉) · R. Katarya
Department of Computer Science and Engineering, Delhi Technological University, Delhi, India
e-mail: aakanshagupta.74@gmail.com

R. Katarya
e-mail: rahulkatarya@dtu.ac.in

© The Author(s), under exclusive license to Springer Nature Singapore Pte Ltd. 2022   111
J. Mathew et al. (eds.), *Responsible Data Science*, Lecture Notes
in Electrical Engineering 940, https://doi.org/10.1007/978-981-19-4453-6_7

predictions. We developed a data-driven epidemiological model for public health crises in this study. We may overcome the limitations of classic epidemiological models that employ only a single component by integrating features from COVID-19-related online information.

To address this issue, we use an LSTM network [11] with a PAN-LDA [12] module in our epidemiological model to revise the infection rate and increase the model's prediction accuracy. This article introduces the SIR-based model, incorporated with the PAN-LDA and the LSTM module for forecasting the COVID-19.

2 Framework of the Model

This paper developed a mathematical model on COVID-19 based on the SIR model and extended it by designing different infection rates. The model network diagram and the interaction among the individual components are shown in Fig. 1. Furthermore, the LSTM method is utilized to optimize the epidemic model's infection rate and is paired with the SIR model to predict the number of infected patients. This research employs the PAN-LDA model to extract features from relevant textual data to analyze the impact of preventive and control measures, transparent reporting on online platforms, and public awareness related to epidemic prevention. The collected characteristics are then coupled with the LSTM method to revise the infection rate deviation calculated by the SIR model. The equations for the SIR model can then be established as

$$\frac{dS(t)}{dt} = -\beta S(t) \tag{1}$$

$$\frac{dI(t)}{dt} = \beta S(t) - rI(t) \tag{2}$$

$$\frac{dR(t)}{dt} = rI(t) \tag{3}$$

where, β and r are the infection rate and recovery rate, respectively.

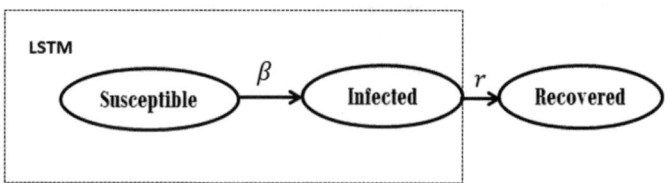

Fig. 1 Flow diagram of the SIR model

3 Prediction of the COVID-19

The traditional epidemiological models use historical data to forecast the infectious disease spread but do not take into account other variables such as media reporting and preventative and control efforts. As a result, known data-driven models must be incorporated to optimize the parameters of epidemic models. The LSTM network is a common method for modeling the hidden variables that are typically used for prediction, such as the number of potentially infected persons. However, experiments have demonstrated that the LSTM model alone cannot accurately estimate the number of infected patients. Given that preventative and control efforts and public knowledge of the disease play an important role in the transmission of infectious diseases, therefore, in this research, we used the PAN-LDA model to extract features from social media posts and news items related to the COVID-19 pandemic. The LSTM network is then embedded with these features, and the number of infected patients is then predicted by updating the infection rate calculated by the classic SIR model. To increase the accuracy of epidemic prediction, the proposed approach updates the infection rate using news information and social media data.

3.1 Textual Data Extraction

We collected the Tweets, Reddit posts, and Google news related to COVID-19 for March 1, 2021, to May 16, 2021. The Tweets are collected from the official Twitter account of the Ministry of Health and Family Welfare, Government of India [13]. Next, we gathered coronavirus-related comments from 8 various subreddits, including r/indiacorona, r/CoronavirusIndia, r/covidIndia, r/COVID19, r/Coronavirus, r/COVID19support, r/nCoV, and r/CoronaVirus2019nCoV. Lastly, we collected English web news from Google search engine using the search terms "covid", "coronavirus", "corona", "India", and "news" to retrieve the news items. Later the duplicates and irrelevant data that did not contain COVID-19 related information were filtered out from all three datasets. The data are classified by date, and the average of all features from the textual data of the day t is presented as a single textual feature vector for that day. The textual data is then pre-processed, and the bag-of-words representation of the textual data on day t is presented as input to the PAN-LDA, which is then converted into feature vectors.

3.2 LSTM Network Based on Textual Features and İnfection Rate

Infection rate-based epidemic models cannot foresee policy changes or emergencies and cannot make short-term predictions. As a result, we introduce the LSTM module

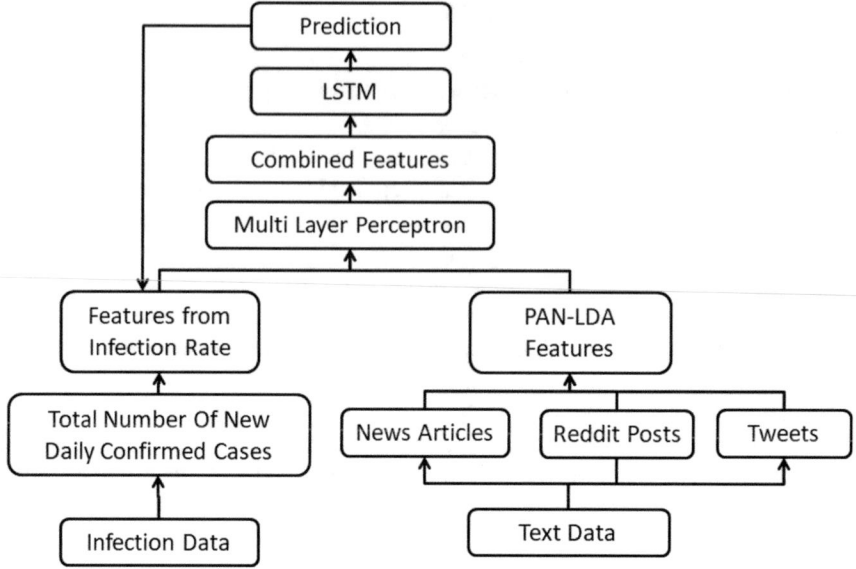

Fig. 2 LSTM network based on textual features and infection rate

based on textual features to mimic social media and existing policy, as illustrated in Fig. 2 to assure short and long-term stability.

In the proposed model, we consider $\beta(t)$ as the real infection rate and that the regressed infection rate using an exponential function is $\hat{\beta}(t)$. The neural network is used to anticipate the difference between the actual and regress infection rates. To account for the influence of news and policy, we mix the textual features established in the preceding section with the infection rate bias. We utilized the LSTM network for encoding hidden states and temporal information. We use a single-layer perception model to convert infection rates and textual features into vectors.

Given infection features $s1$ and textual features $s2$, with $w1$ and $w2$ being the weights of the first two perceptron models. The convolution function $g(.)$ followed by leaky ReLU is as follows:

$$f1 = g(s1; w1) \tag{4}$$

$$f2 = g(s2; w2) \tag{5}$$

The $f1$ and $f2$ are combined into a mixed feature f. Let $f(t)$ be the combined feature at timestamp t, considering $h(t-1)$ as the hidden state at timestamp $t-1$. The following $l()$ includes the fully connected layer and LSTM module that converts the hidden state into prediction. Then

$$(x(t), h(t)) = \mathrm{lstm}(f(t),\ h(t-1); wn) \tag{6}$$

where $x(t)$ is the output, $h(t)$ is the new hidden state, and wn is the network's weight; the Adam optimizer is used as the optimization approach. As the loss function, the mean squared error between prediction and actual is used.

4 Experimental Results

In this section, the proposed model is validated for four Indian states: Maharashtra, Kerala, Karnataka, and Delhi. The number of illnesses handled between March 1, 2021, and May 16, 2021, is utilized as training data to forecast the number of infections between May 17, 2021, and May 24, 2021.

We compared the classic SIR model, SIR + LSTM, and the proposed model to validate the efficacy of the proposed model. We also utilize the proposed epidemic model to analyze the applicability of incorporating data from multiple social media platforms for pandemic prediction. In order to explore the effects of social media data, we compare the SIR + LSTM + features from Twitter and SIR + LSTM + features from Google news, Reddit posts, and Twitter. Tables 1, 2, 3 and 4 indicate the outcomes of the comparison.

As seen in Tables 1, 2, 3 and 4, the proposed model significantly outperforms the classic SIR model. Also, it can be noticed that inclusion of COVID-19 related data from multiple sources has increased the prediction accuracy. The model that incorporates textual information from several online sources outperforms the other models in terms of precision. This discovery demonstrates that language features give extra information and guides for disease prediction.

Table 1 Predicted and actual cases in Maharashtra

Date	SIR	SIR + LSTM	SIR + LSTM + TW	SIR + LSTM + GN + RD + TW	Actuals
17 May	31,691,889	31,276,786.00	31,477,819	31,981,811	31,338,407
18 May	32,976,321	31,868,768.00	31,478,778	31,989,318	31,588,717
19 May	31,246,499	31,978,978.00	31,898,988	31,865,356	31,874,364
20 May	27,086,291	31,676,522.00	31,898,889	32,899,119	32,154,275
21 May	26,908,162	31,786,763.00	31,190,922	32,991,388	32,441,776
22 May	33,798,249	31,786,763.00	31,909,111	32,889,191	32,723,361
23 May	33,036,852	32,878,117.00	31,998,989	32,777,118	33,013,516
24 May	35,209,522	31,878,833.00	32,099,933	32,617,819	33,277,290
R^2	0.3250	0.5686	0.5718	0.7113	
MAPE	5.9039	1.5488	1.8274	1.3205	

Table 2 Predicted and actual cases in Kerala

Date	SIR	SIR + LSTM	SIR + LSTM + TW	SIR + LSTM + GN + RD + TW	Actuals
17 May	18,538,374	17,546,464	18,767,676	18,544,535	18,014,842
18 May	18,172,927	17,766,567	18,672,376	17,888,774	18,149,395
19 May	18,268,519	17,667,656	18,367,637	18,467,643	18,289,940
20 May	18,000,224	17,875,564	18,444,988	18,988,665	18,421,465
21 May	18,313,600	18,877,654	18,177,378	18,878,765	18,555,023
22 May	18,657,837	17,878,667	18,773,677	18,989,898	18,681,051
23 May	19,368,790	17,786,657	18,387,322	19,889,811	18,794,256
24 May	19,540,265	18,797,878	19,838,787	18,985,676	18,881,587
R^2	0.7198	0.5878	0.3493	0.7526	
MAPE	1.6762	2.8641	2.1723	2.2756	

Table 3 Predicted and actual cases in Karnataka

Date	SIR	SIR + LSTM	SIR + LSTM + TW	SIR + LSTM + GN + RD + TW	Actuals
17 May	27,744,501	27,978,176	27,561,655	27,512,454	27,976,933
18 May	28,057,856	27,798,789	27,461,767	27,655,656	28,070,180
19 May	27,764,393	27,089,181	28,287,722	27,674,445	28,199,718
20 May	28,373,754	27,871,117	28,676,655	28,234,356	28,320,429
21 May	28,020,899	27,977,181	28,475,544	28,443,546	28,453,442
22 May	28,235,857	28,889,718	28,564,546	28,554,344	28,582,203
23 May	28,360,876	27,987,917	28,934,118	28,674,677	28,707,320
24 May	28,772,799	28,711,881	28,544,454	28,777,555	28,816,043
R^2	0.8205	0.6154	0.8219	0.9702	
MAPE	0.8369	1.5141	0.8867	0.7103	

5   Conclusion

We developed a SIR-based model with LSTM incorporating features from multiple social media platforms to forecast the trend of the COVID-19. In this study, the PAN-LDA model is used to evaluate and extract COVID-19 related news and awareness information from numerous online sources, which is then encoded into semantic characteristics. The characteristics are then incorporated into the LSTM model to revise the infection rate provided by the SIR model. The model's prediction results are very consistent, demonstrating that the proposed model can forecast infection cases, and textual information processing of related news helps increase the model's accuracy.

Table 4 Predicted and actual cases in Delhi

Date	SIR	SIR + LSTM	SIR + LSTM + TW	SIR + LSTM + GN + RD + TW	Actuals
17 May	18,294,693	17,876,756	18,467,652	18,254,334	18,342,482
18 May	**18,412,220**	17,861,866	18,586,754	18,223,567	18,407,486
19 May	18,478,783	18,678,564	18,345,445	18,635,678	18,474,059
20 May	18,409,091	18,643,378	18,476,532	18,434,677	18,532,803
21 May	17,922,381	18,088,867	18,434,546	18,543,357	18,595,993
22 May	19,007,880	17,866,549	18,544,357	18,585,428	18,659,148
23 May	18,694,718	18,897,776	18,765,544	18,654,954	18,727,191
24 May	19,163,520	19,877,887	19,554,677	18,533,803	18,788,697
R^2	0.6309	0.6760	0.6830	0.7219	
MAPE	1.0798	2.6111	1.0526	0.6630	

Bold number indicates statistical significance

References

1. Bhadra A, Mukherjee A, Sarkar K (2021) Impact of population density on Covid-19 infected and mortality rate in India. Model Earth Syst Environ 7:623–629. https://doi.org/10.1007/s40808-020-00984-7
2. Gupta A, Katarya R (2020) Social media based surveillance systems for healthcare using machine learning: a systematic review. https://europepmc.org/articles/PMC7331523. https://doi.org/10.1016/j.jbi.2020.103500
3. Chan AKM, Nickson CP, Rudolph JW, Lee A, Joynt GM (2020) Social media for rapid knowledge dissemination: early experience from the COVID-19 pandemic. https://www.ncbi.nlm.nih.gov/pmc/articles/PMC7228334/. https://doi.org/10.1111/anae.15057
4. Alessa A, Faezipour M (2018) A review of influenza detection and prediction through social networking sites. 1–27. https://doi.org/10.1186/s12976-017-0074-5
5. Ambikapathy B, Krishnamurthy K (2020) Mathematical modelling to assess the impact of lockdown on COVID-19 transmission in India: model development and validation. JMIR Public Heal Surveill 6:e19368. https://doi.org/10.2196/19368
6. Jia L, Chen W (2021) Uncertain SEIAR model for COVID-19 cases in China. Fuzzy Optim Decis Mak 20:243–259. https://doi.org/10.1007/s10700-020-09341-w
7. Mahajan A, Sivadas NA, Solanki R (2020) An epidemic model SIPHERD and its application for prediction of the spread of COVID-19 infection in India. Chaos Solitons Fractals 140:110156. https://doi.org/10.1016/j.chaos.2020.110156
8. Korolev I (2021) Identification and estimation of the SEIRD epidemic model for COVID-19. J Econom 220:63–85. https://doi.org/10.1016/j.jeconom.2020.07.038
9. Hethcote HW (2007) The mathematics of infectious diseases the mathematics of infectious diseases *. Soc Ind Appl Math 42:599–653
10. Ordun C, Purushotham S, Raff E (2020) Exploratory analysis of Covid-19 tweets using topic modeling, UMAP, and digraphs. arXiv
11. Hochreiter S, Schmidhuber J (1997) Long short-term memory. Neural Comput 9:1735–1780
12. Gupta A, Katarya R (2021) PAN-LDA: a latent dirichlet allocation based novel feature extraction model for COVID-19 data using machine learning. Comput Biol Med 138:104920. https://doi.org/10.1016/j.compbiomed.2021.104920
13. Ministry of Health Management: Ministry of Health. https://twitter.com/MoHFW_INDIA. Accessed 15 Nov 2021

Joint Geometrical and Statistical Alignment Using Triplet Loss for Deep Domain Adaptation

R. Satya Rajendra Singh, Rakesh Kumar Sanodiya, and P. V. Arun

1 Introduction

At present, we encounter many situations where we need to distinguish one person from another. There are many ways to distinguish one person from another, such as fingerprints, DNA, face, and gait [1]. Due to the inexpensive implementation and non-invasiveness of image acquisition, facial recognition is the primary method implemented by cutting-edge classic machine learning and deep learning strategies without the involvement of volunteers. However, several important issues still have to be addressed with these methods. One of the issues is the diversity of the photo shooting environment between the training and test images. For example, in the real world, photographs used in identity cards or electronic passports are taken in special environments, such as pose-c27 (from a central camera with a bright background and various gestures from PIE dataset) and are often used as training images for the classical and deep learning methods. In contrast, we took test images in a volatile environment such as pose-c07 (from a side camera with a dark background and different gestures). Thus, the resulting test images include blur, varying gestures, lights, noise, and arbitrary poses, as shown in Fig. 1, making recognition difficult. If such images are provided to any classifier trained on classical and deep leaning methods, it will encounter two challenges.

The first challenge is the immense distribution gap among the source domain images treated as training images and the test images from the target domain images due to different poses (e.g., c27 and c22). In Fig. 1, although two domains (c27

R. Satya Rajendra Singh · R. K. Sanodiya (✉) · P. V. Arun
Indian Institute of Information Technology, Sri City, Chittor, India
e-mail: rakesh.s@iiits.in

R. Satya Rajendra Singh
e-mail: satyarajendra.rs@iiits.in

P. V. Arun
e-mail: arun.pv@iiits.in

© The Author(s), under exclusive license to Springer Nature Singapore Pte Ltd. 2022
J. Mathew et al. (eds.), *Responsible Data Science*, Lecture Notes
in Electrical Engineering 940, https://doi.org/10.1007/978-981-19-4453-6_8

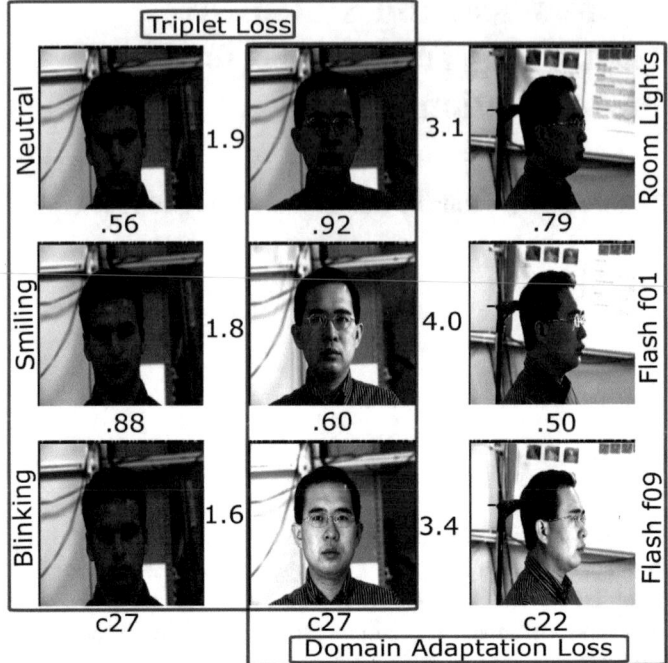

Fig. 1 The distance among the faces of the same person and different people under different face orientations and lighting adjustments is shown in figure

and c22) have images of the same person, they are separated by greater distance than different individuals in the same domain, i.e., the distance between the last two column images of same person in the bottom row is 3.4, which is greater than the distance between the images of two different individual by 1.6. To address this challenge, various transfer learning (TL) or domain adaptation (DA) approaches have been proposed [2, 3].

The second challenge is that the DA methods can only mitigate the distribution deviation among the two domains, but cannot remove the domain discrepancy. Therefore, the hyper-plane learned from the source domain can straightforwardly miscategorize the target domain samples if samples are far from their respective centers or from the edge of their clusters [4]. A general approach is to enforce samples from the target domain with will inter-class separability in addition to intra-class compactness, i.e., samples draw nearer, which are similar, and samples move far away, which are dissimilar. Nevertheless, labeled information for the target domain to preserve both conditions, i.e., inter-class separability together with intra-class compactness, is not available. Fortunately, we have labeled information of the source domain. Recall that both the source domain and the target domain are related. So, the target domain samples also follow the same properties as the source domain samples. Thus, we can use source domain labeled information to preserve both conditions [5]. The existing domain adaptation approaches preserve these conditions by considering any one of the losses such as linear discriminant analysis (LDA) [6], structural risk minimization (SRM) [7], label smoothness consistency (LSC) [8], instance-based discriminative loss, and center-based discriminative loss. According to Schroff et al. and Sun et al. [9, 10], the downside of all these losses are their indirectness and inefficiency for face recognition applications, i.e., they encourage all faces of same person to be projected onto a single point in the embedding space.

In contrast to these losses, Schroff et al. [10] stated that a triplet loss, that is based on large margin nearest neighbor (LMNN) [11], is more effective loss for face recognition applications. The triplet loss attempts [10] to draw a margin between each pair of faces of one person's face and all other faces. Therefore, it allows one person's face to live on manifold, while still enforcing the other person's faces to maintain some distance. Thus, the faces of different individuals can be easily differentiated.

Therefore, it can be concluded from the above reasoning that in order to develop a good face recognition system (or robust classifier), we need to consider sub-centric, data centric, and discriminative losses in an integrated framework [12]. Zhang et al. [6] proposed a method called joint geometrical and statistical alignment (JGSA), that considers all of them into a common framework. However, this framework has some limitations such as it cannot extract complex features due to shallow approach and includes traditional LDA and SA losses. Similarly, the contributions of this paper are summarized as follows.

- We introduce a novel joint geometrical and statistical alignment using the triplet loss (JGSAT) method for deep domain adaptation, that incorporates MMD loss, CORAL loss, and triplet loss into a common framework.
- To show the efficacy of each considered losses in proposed framework, we again introduce two new methods namely triple-MMD (Triple-M) and triple-CORAL (Triple-C).

2 Related Work

There are several methods to overcome the limitations of domain adaptation methods. However, here, we will only discuss methods related to our proposed framework. Based on the way, features are extracted, existing domain adaptation approaches can be classified into two approaches: shallow domain adaptation and deep domain adaptation.

2.1 Shallow Domain Adaptation Approaches

Pan et al. introduced transfer component analysis (TCA) [3] which learns some transfer features across the source and the target domains in reproducing kernel Hilbert space (RKHS) using MMD. Long et al. proposed transfer feature learning with joint domain adaptation (JDA) [13] to enhance the performance of TCA by also taking into account both marginal distribution shift and conditional distribution shift. Fernando et al. proposed a subspace alignment (SA) [14] method which uses some transformation matrix to align source domain vectors with the target domain vectors. In these approaches, to minimize the domain shift, integrated transformation is not required. Nonetheless, the projected variances of the source and target domains are still far apart after applying the SA method due to domain shift. Because of the aforesaid reasons, SA methods are not able to reduce the distribution difference between the source domain and the target domain. When the domain shift is not linear, it does not give good results. Sun et al. overcame the limitations of the SA method by introducing the subspace distribution alignment (SDA) [15] method. The variance of orthogonal principal components is applied in this method. To control the dimensionality of the subspace, all of these domain adaptation methods would require a proper selection of the hyperparameter. To overcome the limitations of SDA, CORAL [16] was introduced by Sun et al. Basically, this method takes into account the eigenvalues and eigenvectors of subspaces of both the source and the target domains without selecting dimensionality of those subspaces. Su et al. proposed a novel method called transductive domain adaptation with affinity learning (TDA-AL) [17] that approaches the problems of domain adaptation tasks in a transductive setting. In this method, a similarity matrix is formed to hold the inter- and intra-domain similarities together. The method then learns the exact similarity across manifold samples. Zang et al. proposed a semi-supervised approach, the semi-supervised transfer discriminant analysis (STDA) [18] method, that overcomes the incompetence of its standard counterpart. Wang et al. improved the performance of the JGSA method by introducing a manifold embedded distribution alignment (MEDA) [7] approach. More specifically, MEDA addresses two challenges:(1) unevaluated distribution alignment and (2) degenerated feature transformation. Luo et al. introduced close yet discriminative DA (CDDA) [8] and discriminative and geometry aware DA (DGA-DA) [19] approaches. CDDA searches a shared feature subspace where both domain data are not only aligned but also discriminative (i.e., instances of different classes are well separated). DGA-DA extends the CDDA approach by considering geometric structures of the data.

2.2 Deep Domain Adaptation Approaches

Tzeng et al. introduced the deep domain confusion method (DDC) [20] with a new CNN architecture which includes an adaptation layer and domain confusion loss to lean domain invariant features. Long et al. introduced a deep domain adaptation network (DAN) [21] approach. The DAN approach generalizes deep CNN to the domain adaptation scenario by embedding hidden representation of all tasks in a reproducing kernel Hilbert space. Sun et al. [22] extended CORAL to lean a nonlinear transformation. Chen et al. proposed a deep domain adaptation framework called joint domain alignment and discriminative feature learning (JDDA) [23], that only considers sub-centric and discriminative losses, but lacks data-centric loss such as maximum mean discrepancy (MMD). More specifically, to preserve the discriminative information of the source domain, JDDA considers instance-based method (JDDA-I) and center-based method (JDDA-C).

3 Proposed Method: Joint Geometrical and Statistical Alignment Using Triplet Loss

In this section, we explain the joint geometrical and statistical alignment using the triplet loss method in detail. Following the previous work of Chen et al. [23], we adopt two-stream CNN architecture with shared weights. The first stream takes the source domain image as input, while the second stream takes the target domain image. Our work is different from other shallow or deep domain adaptation approaches in the way that we consider a robust triplet loss function as a discriminant loss and reduce the distribution gap statistically and geometrically in deep learning frameworks. This encourages domain invariant features for better final classification.

Following unsupervised DA settings, we define source domain labeled data as $\mathcal{D}_s = \{X_s, Y_s\} = \{(x_j, y_j)\}_{j=1}^{n_s}$ and target domain unlabeled data as $\mathcal{D}_t = \{X_t\} = \{(x_j)\}_{j=1}^{n_t}$, where both domain samples have the same dimensions, i.e., $x \in \mathbb{R}^d$. Let us assume that Θ is a shared parameter in both domains to be learned. $F_s \in \mathbb{R}^{B \times N}$ and $F_t \in \mathbb{R}^{B \times N}$ represent learned deep features from the bottleneck layer with respect to the source domain stream and the target domain stream, respectively. N and B represent the number of hidden neurons in the bottleneck layer and batch size during the training stage, respectively. Then, the network of proposed model can be trained by using the following loss function.

$$\mathcal{L}(\Theta|X_s, Y_s, X_t) = \mathcal{L}_s + \alpha\mathcal{L}_t + \beta\mathcal{L}_m + \gamma\mathcal{L}_c \tag{1}$$

$$\mathcal{L}_s = \frac{1}{n_s} \sum_{i=1}^{n_s} C(\Theta|x_s^i, y_s^i) \tag{2}$$

$$\mathcal{L}_t = \mathcal{T}_d(\Theta|X_s, Y_s) \tag{3}$$

$$\mathcal{L}_m = \mathrm{MMD}(F_s, F_t) \tag{4}$$

$$\mathcal{L}_c = \mathrm{CORAL}(F_s, F_t) \tag{5}$$

Here, $\mathcal{L}_s$, $\mathcal{L}_t$, $\mathcal{L}_m$, and $\mathcal{L}_c$ represent source domain loss, triplet loss, MMD loss, and CORAL loss, respectively. α, β, and γ are trade-off parameters to balance the contribution of the discriminative loss, the MMD loss and the CORAL loss. Particularly, $C(\Theta|x_s^i, y_s^i)$ represents the source domain classification loss. $\mathcal{T}_d(\Theta|X_s, Y_s)$ represents source domain discriminative loss, which ensures better inter-class separability and intra-class compactness. $\mathcal{L}_m = \text{MMD}(F_s, F_t)$ and $\mathcal{L}_c = \text{CORAL}(F_s, F_t)$ represent domain discriminative losses measured by MMD and CORAL, respectively.

3.1 Triplet Loss

For making the two-stream CNN network to learn even more discriminative features, we apply triplet loss that directly demonstrates what we want to achieve in classification, validation, and clustering. Motivation behind considering this loss is that it prompts all faces of one subject to be projected onto a single point in the embedding space. However, at the same time, it tries to draw a margin between each pair of faces of one subject and all other faces.

In triplet loss [10], the distance between an image x_i^a (anchor) of a specific subject to all other images x_i^p (positive) of the same subject is lesser than that to any image x_i^n (negative) of any other subject.

The triplet loss function can be defined as follows:

$$\left\| f(x_j^a) - f(x_j^p) \right\|_2^2 + \Delta < \left\| f(x_j^a) - f(x_j^n) \right\|_2^2, \forall (f(x_j^a), f(x_j^p), f(x_j^n)) \in \mathcal{T} \quad (6)$$

where $\mathcal{T}$ with cardinality M is a set of all possible triplets in the training. α is a margin that is drawn between the positive and the negative pairs. The loss that is being minimized is then

$$\mathcal{T}_d = \sum_i^M [\left\| f(x_s^a) - f(x_s^p) \right\|_2^2 - \left\| f(x_s^a) - f(x_s^n) \right\|_2^2 + \Delta]_+ \quad (7)$$

3.2 Distribution Divergence Minimization

Since, in the target domain, there is little or no label data available, only training the classifier with the source domain label data leads to over-fitting to the source distribution. As a result, the trained classifier performs poorly for the target domain. Our intuition is that if we are able to learn a representation where the distribution difference between the source domain and the target domain data is minimal, we can avoid over-fitting the classifier that leads to a minimal loss in accuracy for the target domain.

To minimize the distribution difference between the source and the target domains data, we consider the standard distribution distance metric called maximum mean discrepancy (MMD) [3]. In MMD, the distance between two domains can be computed as follows:

$$\text{MMD}(F_s, F_t) = \left\| \frac{1}{|F_s|} F_s - \frac{1}{|F_t|} F_t \right\|_F^2 \tag{8}$$

where $\|\cdot\|_F^2$ is the squared matrix Frobenius norm and $|\cdot|$ is the number of elements in $(\cdot)$.

3.3 Correlation Alignment

In subspace alignment methods, domain shift is minimized by aligning the covariance of the features in source and target domains. Therefore, the domain discrepancy loss calculated by CORAL can be expressed as

$$L_c = \text{CORAL}(F_s, F_t) = \frac{1}{4L^2} \|\text{Cov}(F_s) - \text{Cov}(F_t)\|_F^2 \tag{9}$$

where $\|\cdot\|_F^2$ denotes the squared matrix Frobenius norm. $\text{Cov}(F_s)$ and $\text{Cov}(F_t)$ denote the covariance matrices of the source and the target features in the bottleneck layer, which can be calculated as $\text{Cov}(F_s) = F_s^T J_b F_s$ and $\text{Cov}(F_t) = F_t^T J_b F_t$. $J_b = I_b - \frac{1}{b} 1_n 1_n^T$ is the centralized matrix, where $1_b \in \mathbb{R}^b$ is an all-one column vector, and b is the batch size.

3.4 Training

The proposed method can easily be implemented via mini-batch SGD [23]. For this, the total loss is given as $\mathcal{L} = \mathcal{L}_s + \alpha \mathcal{L}_t + \beta \mathcal{L}_m + \gamma \mathcal{L}_c$, while the source loss is defined by the conventional softmax classifier. $\mathcal{L}_t$ defined in (7), $\mathcal{L}_m$ defined in (8), and $\mathcal{L}_c$ defined in (9) are differentiable with respect to the inputs. Therefore, the parameters Θ can be updated by the standard back propagation directly.

$$\Theta_{t+1} = \Theta_t - \eta \frac{\partial(\mathcal{L}_s + \alpha \mathcal{L}_t + \beta \mathcal{L}_m + \gamma \mathcal{L}_c)}{\partial(x_i)} \tag{10}$$

where η is the learning rate.

4 Experiments

In this section, we conduct experiments on PIE face dataset to verify the effectiveness of our proposed methods (Triplet-M, Triplet-C, and JGSAT). We compare our proposed methods with several state-of-the-art shallow domain adaptation methods such as TDA-AL, MEDA, CDDA, DGA-DA, STDA, and JGSA, and deep domain adaptation methods such as DDC, DAN, Deep-CORAL, JDDA-I, and JDDA-C.

4.1 Setup and Implementation Details

PIE, which stances for "pose, illumination, expression", is used because of its extensive use in the past by many others state-of-the-art domain adaptation algorithms to show the comparisons. The PIE face recognition dataset contains images of 68 subjects under 21 different illumination conditions. The images in the dataset have been assembled into five different groups containing different poses, namely PIE-C05, PIE-C07, PIE-C09, PIE-C27, and PIE-C29 where C05 indicates left pose, C07 indicates upward pose, C09 indicates downward pose, C27 indicates frontal pose, and C29 indicates right pose. One of these pose is selected as the source domain and another as the target domain. For example, 5→7, here five represents pose source dataset C05 while seven represents pose target dataset C07. The source domain pose dataset 5 labeled information is used for improving the performance of target domain pose dataset 7 classifier.

To analyze the effectiveness of our proposed approaches and other deep domain adaptation approaches on PIE datasets tasks, we use modified LeNet. For a fair comparison, all the deep learning-based domain adaptation models as discussed above have the same architecture as our approaches. Note that all of the above approaches are implemented through Tensorflow and trained with the Adam optimizer with learning rate $\eta = 10^{-4}$. The parameter values of JDDA and other methods are kept as reported in their paper.

To exhibit the effectiveness of incorporating two domain losses in our proposed method JGSAT (Triplet-MC) loss function, we neutralize the effect of one domain loss function at a time and run the experiment. As a result, we obtain two more methods called Triple-M and Triple-C, where in Triple-M method we neutralize CORAL loss, but in Triple-C method we neutralize MMD loss.

Since the performance of any method depends on an appropriate value of each parameter, we performed parameter sensitivity testing by keeping the other parameters constant and changing one parameter value at a time. After conducting the parameter sensitivity test, we find that proposed method perform well for the parameter values $\alpha = 0.001$, $\beta \in [10^{-3}, 10^{0}]$, and $\gamma \in [10^{-3}, 10^{0}]$.

5 Results and Discussion

The results acquired by envisioned techniques JGSAT (Triplet-MC), Triplet-M, and Triple-C, and other deep domain adaptation approaches after exploring on all potential tasks of the PIE face dataset, the results obtained by our projected systems are reported in Table 1. However, the results of the shallow domain adaptation approaches are taken directly from their respective papers. In addition, we have also plotted a graph as shown in Fig. 2 showing the variation of accuracy over iterations of the deep domain adaptation approaches for the task $5\rightarrow09$.

From Table 1, it can be seen that the deep DA methods have an edge over the shallow methods in performance because of extracting the deep features rather than considering the handcrafted features. JGSAT (Triplet-MC) outperforms all other comparative domain adaptation approaches on tasks $5\rightarrow7$, $5\rightarrow27$, $5\rightarrow29$, $9\rightarrow29$, $27\rightarrow5$, $27\rightarrow7$, $27\rightarrow9$, and $27\rightarrow29$. Similarly, other proposed method Triplet-M method outperforms for tasks $5\rightarrow9$, $9\rightarrow7$, and $29\rightarrow27$. However, STDA outperforms for three tasks $7\rightarrow5$, $7\rightarrow27$, and $29\rightarrow5$.

Comparing the average performance on all tasks of PIE dataset, our envisioned approach "JGSAT" prevailed over the second-best domain adaptation method "DAN" by 6.01%. Although the shallow DA method JGSA performs both geometrical and statistical alignments, our envisioned approach "JGSAT" has a 55% improvement. This is because of extracting deep features instead of handcrafted features. The superior performance of Triplet-C compared to JDDA-I and JADA-C strongly suggests that it is noteworthy to preserve discriminatory information using triplet loss rather than center-based or instance-based methods. The attainment of DAN has improved over DDC due to multi-kernel MMD. After comparing our own proposed methods (i.e., Triple-M, Triple-C, and JGSAT), it can be seen that it is valuable to consider both CORAL and MMD losses to obtain better results. The exploits of associated deep DA methods over the number of iterations have shown in Fig. 2, can see that

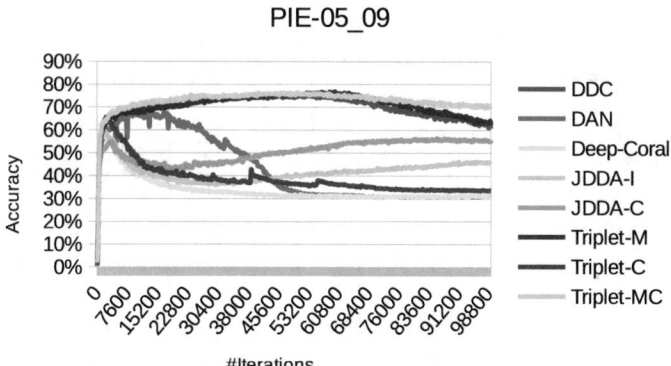

Fig. 2 Plot showing variation of accuracies over iterations for the deep domain adaptation approaches

Table 1 Accuracy (%) on PIE face dataset

Tasks	Shallow domain adaptation approaches					STDA [18]	Deep domain adaptation approaches					Triplet-M	Triplet-C	JGSAT (Triplet-MC)
	TDA-AL [17]	MEDA [7]	CDDA [8]	DGA-DA [8]	JGSA [6]		DDC [20]	DAN [21]	Deep-CORAL [22]	JDDA-I [23]	JDDA-C [23]			
PIE 1 5→7	35.97	49.60	60.22	65.32	68.07	67.22	69.16	60.91	51.59	49.79	53.33	68.19	60.80	**70.73**
PIE 2 5→9	32.97	48.40	58.70	62.81	67.52	68.57	75.41	67.77	58.40	59.72	56.45	**76.73**	65.48	76.45
PIE 3 5→27	35.24	77.23	83.48	83.54	82.87	90.03	99.72	99.82	**99.86**	99.82	**99.86**	99.75	**99.86**	**99.86**
PIE 4 5→29	28.43	39.82	54.17	56.07	46.50	57.35	68.47	54.23	45.06	52.22	48.95	70.84	50.41	**71.59**
PIE 5 7→5	38.90	58.49	62.33	63.69	25.21	**66.00**	45.52	41.38	37.01	39.13	29.30	45.31	39.09	**46.14**
PIE 6 7→9	49.39	55.27	64.64	61.27	54.77	67.70	70.90	96.11	88.88	89.86	85.27	79.23	87.98	90.34
PIE 7 7→27	53.26	81.25	79.90	82.37	58.96	**83.87**	77.22	49.88	55.79	58.57	61.70	77.88	61.70	75.83
PIE 8 7→29	36.95	44.05	44.00	46.63	35.41	50.86	73.19	98.88	88.68	93.75	92.01	71.25	91.87	93.95
PIE 9 9→5	34.03	56.24	58.46	56.72	22.81	60.32	77.29	**97.98**	95.41	95.97	91.38	76.18	95.20	95.60
PIE 10 9→7	49.54	57.82	59.73	61.26	44.19	58.69	67.01	60.79	53.43	72.01	44.89	**67.84**	56.31	65.90
PIE 11 9→27	48.99	76.23	77.20	77.83	56.86	77.44	85.34	86.14	72.63	72.08	70.41	86.56	77.53	84.86
PIE 12 9→29	39.34	53.06	47.24	44.24	41.36	50.55	84.79	96.45	96.45	94.16	96.31	80.20	96.04	**97.22**
PIE 13 27→5	42.20	85.95	83.10	81.84	72.14	88.57	99.68	99.90	99.89	99.93	99.93	99.51	99.92	**100**
PIE 14 27→7	63.90	78.20	82.26	85.27	88.27	89.38	90.27	83.61	75.20	77.98	72.98	89.30	76.87	**92.36**
PIE 15 27→9	61.64	80.20	86.64	90.95	86.09	89.09	90.13	87.36	84.23	81.66	83.26	89.79	85.34	**91.59**
PIE 16 27→29	46.32	67.70	58.33	53.80	74.32	71.63	83.61	77.36	67.22	70.55	70.31	83.61	76.59	**84.02**
PIE 17 29→5	32.92	57.71	48.02	57.44	17.52	**60.74**	54.79	46.56	46.14	42.32	39.51	55.48	48.02	**58.43**
PIE 18 29→7	37.26	49.66	45.61	53.84	41.06	53.71	80.90	97.91	97.84	96.59	91.73	74.30	91.59	95.62
PIE 19 29→9	36.64	62.13	52.02	55.27	49.20	62.50	81.94	99.23	96.45	96.87	95.27	83.95	95.55	95.34
PIE 20 29→27	38.96	72.15	55.99	61.82	34.75	69.42	74.68	64.93	52.08	59.79	46.94	**77.50**	61.59	76.18
Average	42.14	62.56	63.10	65.09	53.39	69.42	77.51	78.36	73.11	75.13	71.48	77.67	75.88	**83.10**

Bold face values are best results

the performance of both Triplet-M and JGSAT (Triplet-MC) is better as iteration numbers increases.

6 Conclusion and Future Work

In present treatise, showcased a joint geometrical and statistical alignment using the triplet loss (JGSAT) method for deep domain adaptation reduces domain shift between domains statistically and geometrically simultaneously by incorporating maximum mean discrepancy(MMD), CORAL, and triplet Loss in a unified framework. Extensive experiments and above discussion have verified the inclusion of a deep learning framework for extracting deep features, triplet loss for preserving discriminative information, MMD loss for statistical alignment, and CORAL loss for geometric alignment is noteworthy for achieving better results. Therefore, the exploit of our projected method is ample improved than that of shallow domain adaptation and deep domain adaptation methods.

Considering a real-world problem where voluminous source domains can simultaneously contribute to advancing the pursuance of the target domain. In the subsequent time, we are going to extend our framework to multi-source domain adaption methods by considering more than two CNN stream architectures.

References

1. Hafiz F, Shafie AA, Mustafah YM (2012) Face recognition from single sample per person by learning of generic discriminant vectors. Procedia Eng 41:465–472
2. Sanodiya RK, Mathew J, Paul B, Jose BA (2019) A kernelized unified framework for domain adaptation. IEEE Access 7:181381–181395
3. Pan SJ, Tsang IW, Kwok JT, Yang Q (2011) Domain adaptation via transfer component analysis. IEEE Trans Neural Netw 22(2):199–210
4. Sanodiya RK, Yao L (2021) Discriminative information preservation: a general framework for unsupervised visual domain adaptation. Knowl-Based Syst 227:107158
5. Sanodiya RK, Mathew J (2019) A novel unsupervised globality-locality preserving projections in transfer learning. Image Vision Comput 90:103802
6. Zhang J, Li W, Ogunbona P (2017) Joint geometrical and statistical alignment for visual domain adaptation. arXiv preprint arXiv:1705.05498
7. Wang J, Feng W, Chen Y, Yu H, Huang M, Yu PS (2018) Visual domain adaptation with manifold embedded distribution alignment. In: 2018 ACM multimedia conference on multimedia conference. ACM, pp 402–410
8. Luo L, Chen L, Hu S, Lu Y, Wang X (2017) Discriminative and geometry aware unsupervised domain adaptation. arXiv preprint arXiv:1712.10042
9. Sun, Y Chen Y, Wang X, Tang X (2014) Deep learning face representation by joint identification-verification. In: Advances in neural information processing systems, pp 1988–1996
10. Schroff F, Kalenichenko D, Philbin J (2015) Facenet: a unified embedding for face recognition and clustering. In: Proceedings of the IEEE conference on computer vision and pattern recognition, pp 815–823

11. Weinberger KQ, Saul LK (2009) Distance metric learning for large margin nearest neighbor classification. J Mach Learn Res 10:207–244
12. Sanodiya RK, Mathew J, Saha S, Tripathi P (2020) Particle swarm optimization based parameter selection technique for unsupervised discriminant analysis in transfer learning framework. Appl Intell 50(10):3071–3089
13. Long M, Wang J, Ding G, Sun J, Yu PS (2013) Transfer feature learning with joint distribution adaptation. In: Proceedings of the IEEE international conference on computer vision, pp 2200–2207
14. Fernando B, Habrard A, Sebban M, Tuytelaars T (2013) Unsupervised visual domain adaptation using subspace alignment. In: Proceedings of the IEEE international conference on computer vision, pp 2960–2967
15. Sun B, Saenko K (2015) Subspace distribution alignment for unsupervised domain adaptation. In: BMVC, vol 4, pp 24.1–24.10
16. Sun B, Feng J, Saenko K (2016) Return of frustratingly easy domain adaptation. In: Thirtieth AAAI conference on artificial intelligence
17. Shu L, Latecki LJ (2015) Transductive domain adaptation with affinity learning. In: Proceedings of the 24th ACM international on conference on information and knowledge management. ACM, pp 1903–1906
18. Zang S, Cheng Y, Wang X, Yu Q (2018) Semi-supervised transfer discriminant analysis based on cross-domain mean constraint. Artif Intell Rev 49(4):581–595
19. Luo L, Chen L, Hu S, Lu Y, Wang X (2020) Discriminative and geometry-aware unsupervised domain adaptation. IEEE Trans Cybern 50(9):3914–3927
20. Tzeng E, Hoffman J, Zhang N, Saenko K, Darrell T (2014) Deep domain confusion: maximizing for domain invariance. arXiv preprint arXiv:1412.3474
21. Long M, Cao Y, Wang J, Jordan MI (2015) Learning transferable features with deep adaptation networks. arXiv preprint arXiv:1502.02791
22. Sun B, Saenko K (2016) Deep coral: correlation alignment for deep domain adaptation. In: European conference on computer vision. Springer, pp 443–450
23. Chen C, Chen Z, Jiang B, Jin X (2019) Joint domain alignment and discriminative feature learning for unsupervised deep domain adaptation. Proceedings of the AAAI conference on artificial intelligence 33:3296–3303

Virtual Try-On Using Style Transfer

Ravi Ranjan Prasad Karn, Rakesh Kumar Sanodiya, Eswara Surya Chandaluri, S. Suryavardan, L Ranajith Reddy, and Leehter Yao

1 Introduction

The new age of technology has moved the world to our personal devices and people can now buy clothing from the comfort of their homes. Especially in recent times, there has been a sizeable increase in the number of customers using these online stores. The one drawback though, is the need for trial or try-out rooms. We like to choose a particular clothing item based on how we look when wearing it. Wouldn't it be convenient if we could get a picture of ourselves wearing our desired clothes. This will also aid in the reduction of return rates by narrowing the gap between anticipation and reality.

In this work, we aim to achieve Virtual Try-on by exchange clothing information between given set of images. This would require a joint extraction of the clothing, shape, size and body pose of input images. A primary issue is the lack of an "ideal" dataset. It's easier for a model to learn by comparing its outputs with a ground truth image but in this task we do not have images of people with different clothing. This has inspired the use of encoders and segmentation [1]. Other approaches can also be found in the work on the Person Re-identification task [2]. The images from virtual try-on are used to train a classifier in that case. We have elaborated this in the next section.

We use generative convolutional neural networks to implement virtual try-on by using style transfer. Style transfer in our context refers to the changing of the clothes of a person in an image and our view is restricted to only the person-clothing domain. For example, consider an image A as the input and an image B as the target image. Our objective is to generate an image A' such that it contains the individual in image

R. R. P. Karn (✉) · R. K. Sanodiya · E. S. Chandaluri · S. Suryavardan · L. R. Reddy
Indian Institute of Information Technology, Sri City, Chittor, India
e-mail: raviranjanprasad.k@iiits.in

L. Yao
National Taipei University of Technology, Taipei 10608, Taiwan

© The Author(s), under exclusive license to Springer Nature Singapore Pte Ltd. 2022   131
J. Mathew et al. (eds.), *Responsible Data Science*, Lecture Notes
in Electrical Engineering 940, https://doi.org/10.1007/978-981-19-4453-6_9

B wearing the same clothes as the one in image A. The clothing information from image A and structural information from image B can be encoded. They can then be combined to generate A'. Using such pairs of input and target images, a photo-realistic picture of the target person wearing the desired clothing can be generated. This task will be simpler if a picture containing just the clothing in a plain background or a 3D scan of the garment is available . While the outputs might be of high fidelity, generating a 3D representation or extracting a garment [3] separately can be time-consuming and expensive. This approach is not suitable for real-world applications.

Therefore, we propose two end-to-end models with less architectural complexity and data requirement, each with its own set of merits and demerits that perform style transfer across input images and generate decoded images. One is an Encoder-Decoder based approach and the other uses a Generative Adversarial Network (GAN). While GAN's are becoming increasingly popular, we also cast some light on the ability of auto-encoders in the style transfer domain.

2   Related Work

Virtual Try-On or Cloth Swapping has seen various approaches in recent times. The major approaches can be broadly classified into two methods: **Masking** and **Encoding**. There are significant works done in the field of **Human Parsing and Segmentation** and their inclusion in different applications. Amit Raj et al. [1] worked on segmentation of images which returns a mask which helps in extracting specific and required outputs. Look into Person(LIP) [3] is one such method which uses masking to parse the clothing item from images . While it is a data-centric approach, with an ideal mask, the model can be highly adaptable to different image styles. There are many other methods which focus on body masking to identify the targets and replace their clothing. Extracting the target can be resource intensive but it allows reliable cloth swap in the presence of a clothing-human image pair [4]. An extension to this approach is the use of instance mapping, which makes multi-instance translation easier [5]. Similarly, jointly training clothes and 3D body shapes to tackle cloth swap are effective but require additional information such as depth scans for every target [6]. Other attempts at this problem rely on the existence of a database of usable cloth templates and body models or extracted masks [7].

Research on **Person Re-identification** focuses on identifying a person of interest from different images across multiple cameras. One methodology for this task aims to train the classifier on a generated dataset [2]. Here the generated data is obtained by encoding input images to swap clothing and as a result, add diversity to the data. In our proposed work, We do not require the classifier but our generative module can be similar to such re-id models.

Auto-Encoders and **GANs** have shown impressive results on image generation tasks such as faces [8], birds [9], image-translation [10] etc. Further development has helped in utilising extra information and thus, generating desired outputs by implementing conditional models. For our problem, the existing methods either have a segmentation stage [1] or depend on having a usable garment dataset [4]. Whereas, our approach implements encoders with a requirement of just human image pairs and relatively less computation.

3 Proposed Work

Our proposed work is based on two models Encoder-decoder-based model and GAN-based model.

3.1 Encoder-Decoder Based Approach

In this approach, the architecture is made up of three major components: Pose Encoder, Visual Encoder and Decoder. The pose of an image relates to the geometrical characteristics of the image such as structure of the face of a person(such as oval, round, square, diamond, etc.), pose in which a person is standing, etc. The visual component of an image relates to the color of the image such as the color of the shirt.

The pose encoder receives an image as input and generates the pose code. All structural characteristics of the input image are represented by this encoding. Similarly, the visual encoder receives image as input and encodes the visual aspect of it to visual code. The decoder takes both the pose and visual code and constructs an image based on these encodings.

Let P represent the pose encoder, V the visual encoder, D the decoder , and X_{gray} the gray-scale version of image X. To perform the style transfer between two images X and Y, the pose and visual codes of X are $P(X_{gray} = X_p$ (pose code) and $V(X) = X_v$ (visual code). The pose encoder takes gray-scale image but not the original(color) image because visual information is not required. Similarly for image Y, they are Y_p and Y_v. The output images (let's call them E and F) can be generated by passing the visual code of X and pose code of Y and vice versa. $E = D(X_v, Y_p)$, $F = D(X_p, Y_v)$

Reconstruction loss is applied between the input image and the output image using a distance metric (Fig. 1).

$$L_{rec} = \|X - D(P(X_{gray}), V(X))\| \tag{1}$$

This ensures reconstruction from input encodings but it does not guarantee that the Visual encoder encodes only the visual characteristics of the image. We need just the clothing but the visual encoder might encode geometrical aspects of the image. To avoid this, we define another loss function (Fig. 2).

$$L_{rec\_gray} = \|X_{gray} - D(P(X_{gray}), null)\| \tag{2}$$

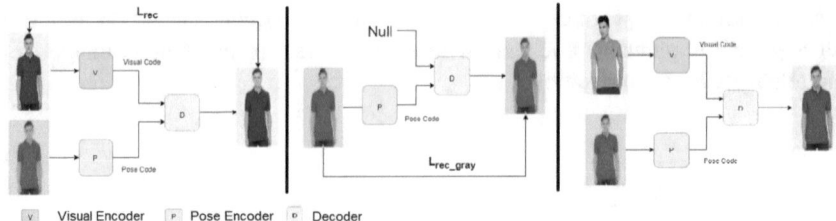

V Visual Encoder   P Pose Encoder   D Decoder

Fig. 1 An overview of the Encoder-Decoder based approach 3.1

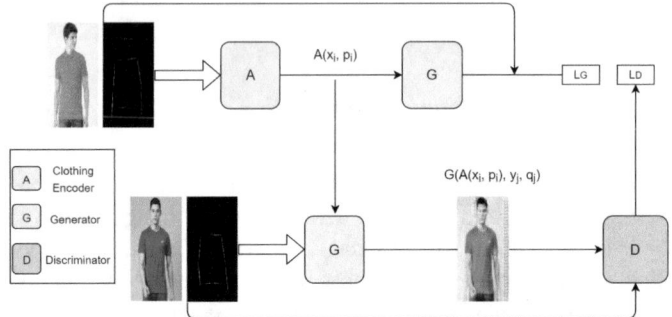

Fig. 2 An overview of the GAN based approach 3.2. L_G is the reconstruction loss from the first stage of the model and L_D is the classification loss from the second stage.

Here the decoder was only provided the pose code and not the visual code. So the decoder constructs an image without having any visual information. Thus, this loss function guarantees that the structure encoder encodes the complete image pose leaving no room for the visual encoder to encode the pose of image.

We train the structure encoder, visual encoder and decoder together using a sum of the losses:

$$L_{\text{Total}} = L_{\text{rec}} + L_{\text{rec\_gray}} \tag{3}$$

3.2 GAN based Approach

This approach can be divided into two stages. We start with the Clothing Encoder which uses the input image and then the encoded output with the target image is passed to the Generator. The generated image and the ground truth or input image is used to train the discriminator. In addition to the input and target images, their poses are also used. The poses were detected using OpenPose [11] and this led to considerable improvements in the fidelity of the outputs.

Reconstruction is the first stage. In this approach, at each epoch, two sets of images and poses are passed as inputs—Input image and Target image. Thus, at this stage the input image with it's pose is provided to the Clothing Encoder. This is a relatively smaller model with 2D convolutional blocks. A single residual block is utilised which includes a residual connection with an instance normalisation layer. A grayscale form of the same input image and pose is now given as input to the generator along with the encoded clothing. Generation or decoding uses a deeper model with 5 residual blocks. Both models use Reflection Padding because the images transition "smoothly" with the re-use of data as it keeps the output at hand with the original distribution.

Transferring is the second stage. The encoded clothing from the previous stage is fed back to the generator. This time, it is accompanied by the target image and it's pose. The image derived at this stage is treated as the final output from this approach. The quality of the image is monitored by the discriminator. It prevents the generator from learning to output the image passed in to the clothing encoder in the first stage. The discriminator is also a deep model with convolutional blocks and a sigmoid activation layer at the end. In this method, instance normalization is used to simplify the learning process. Intuitively, the normalization process removes the instance-specific contrasting information from the content image. This simplifies generation, which is especially beneficial in a task like style transfer.

Optimization. The GAN-based approach includes three models and all models use the Adam optimizer with the same parameters.

The reconstruction loss from the first stage is:

$$L_G = E[\|x_i - G(A(x_i, p_i), x_i, p_i)\|_1] \qquad (4)$$

where x_i is the input image, p_i the input pose, A is encoder and G is the generator. The two models together must be able to consistently encode clothing and output realistic style-transferred images.

At the second stage, the discriminator loss is:

$$L_D = E[\| \log(D(y_i) + \log(1 - D(G(A(x_i, p_i), y_i, q_i))\|] \qquad (5)$$

where y_i is the target image, q_i the target pose and D is the discriminator. Here, we exploit the minimax game [12] and improve the performance of the generative models.

The discriminator uses the binary cross-entropy loss from stage 2. While, the generator and encoder are jointly trained with a sum of the reconstruction loss from stage 1 and the minimax loss derived from stage 2 (Fig. 3).

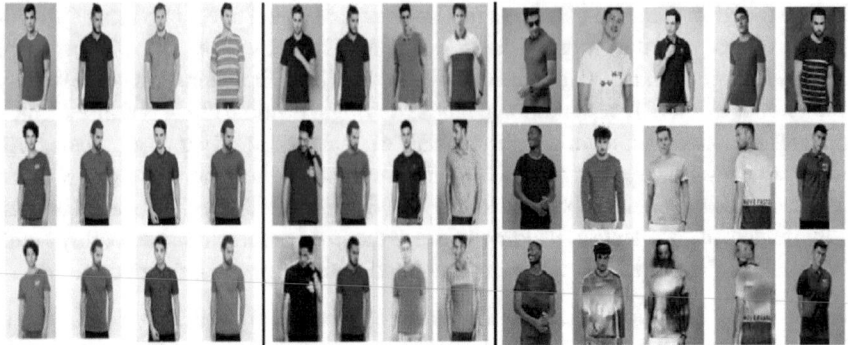

Fig. 3 Left: few outputs from the encoder-decoder based approach 3.1. Middle: some outputs from the GAN based approach 3.2. Right: some outputs of the DG-Net model. The first row is the input, the second is the target and the third is the resultant image output. [2]

4 Experiments

The experiments are done by considering appropriate datasets and implementation design.

4.1 Datasets

As mentioned in the sections above, our model uses only a set of image pairs to perform the style transfer. The data used for training and evaluation does not require masks or separate garments. Thus, we use the Market-1501 dataset and another dataset of 6458 images scraped from a fashion e-commerce website, Myntra. The motive behind using both these datasets was to ensure the images were generalised and varied. This helped in testing the capability of the model on relatively minimal computation and resource utilisation and comparing the results on these two datasets.

4.2 Implementation

The implementation was carried out in Tensorflow framework. **Approach 3.1.** The Visual encoder is fed with the actual image while the pose encoder is fed with a gray-scale representation of the original image (Input image dimensions are $128 \times 64 \times 3$ and shifted to range $[-1, 1]$). Here we refer to convolution block as a combination of the convolution layer, instance normalization and LeakyReLU layers. A series of convolution blocks with one residual connection (in between) makeup the visual encoder. Its output is a one dimensional tensor of length 512. The pose encoder is similar to the visual encoder but it contains additional residual connections

(3 connections) and produces a three dimensional tensor as output ($16 \times 32 \times 8$) (believing that one-dimensional tensor cannot encode structural attributes well). The decoder takes both the visual and the pose code and concatenates them after passing them through a transpose convolution block. The concatenated tensor is passed through series of transpose convolution blocks and the output dimension is the same as the input image dimension.

Approach 3.2. For the GAN based approach, before the data is loaded, we use the OpenPose library to extract the poses and save them. OpenPose jointly detects the human body and structure keypoints in single images. The images with their poses are then loaded using tensorflow data library where each image is resized to $128 \times 64 \times 3$ and it is shifted to range $[-1, 1]$. The data is passed to the models in pairs of input and target images, and each image and its pose is concatenated. The convolutional blocks are made of conv2d layers, instance normalisation and ReLU activation. The same layers but with residual or skip connections form the residual block. Unlike the auto-encoders, the GAN model includes reflection padding in the convolutional blocks. The output from the clothing encoder is of size $32 \times 16 \times 256$ and it is passed to the generator model. The generator's result after passing the target image and pose through the residual blocks is concatenated with the encoded clothing. Then it is connected to transpose convolutional blocks. All these blocks use ReLU activation except for the final layer, which uses Tanh, allowing quick saturation and coverage of the color space of the training distribution. As expected, the generator output is of size $128 \times 64 \times 3$. Finally, the discriminator includes convolutional blocks with LeakyReLU and alpha equal to 0.2. The final layer is a sigmoid layer as it has to predict whether the image is real or fake. The optimizer used for both the approaches is the Adam optimizer with learning rate $2e - 4$ and $(\beta_1, \beta_2) = (0.5, 0.999)$.

4.3 Results

GANs have shown a lot of potential and have been adapted to various tasks but many of those tasks are difficult to evaluate. This is one of the drawbacks both of the above stated approaches, as a human level performance measure is imperative in deep models. "Generative Adversarial Networks are a promising class of models that is often held back by unstable training and by the lack of a proper evaluation metric—as stated in [13].

The **Results** shown are for both our approaches along with the output on our scraped dataset of the DG-Net model pre-trained on market-1501. Despite the lack of a true evaluation metric in such taks, we derive qualitative measures for the generated images. As Virtual-Try On uses input images, using Fretchet Inception Distance (FID) [14] and Structure Similarity Index (SSIM) [15] should help collate and evaluate the appearance and structural encoding. We have provided and compared the input and output similarity results for the models in Table 1.

Table 1 FID and SSIM scores

Metric		DG-Net	Encoder-decoder	GAN
FID	Appearance image	**140.34**	152.47	178.88
	Structure image	170.31	158.76	**156.38**
SSIM	Appearance image	0.752	**0.910**	0.568
	Structure image	0.374	**0.907**	0.588

Bold faced values are best results

5   Conclusion and Future Work

In this work, we propose a style transfer approach to Virtual Try-On using two methods - Encoder-Decoder based and a GAN based which gives satisfactory results on style transfer using just two images without having a comparable ground truth image. Both models are relatively shallow and require minimal computation. After being trained, the model can be used for Virtual Try-On on any input image.

As the resultant images are not as clear as desired which is an inherent drawback of shallow models. Thus, in future, we plan to experiment with segmentation and we will attempt to optimize our model to reduce complexity. We will also explore our options to address issues caused by the lack of an ideal dataset. We are also planning to use transfer learning [16–18] in this area.

References

1. Raj A, Sangkloy P, Chang H, Hays J, Ceylan D, Lu J (2018) Swapnet: image based garment transfer. In: Ferrari V, Hebert M, Sminchisescu C, Weiss Y (eds) Computer vision—ECCV 2018. Springer International Publishing, Cham, pp 679–695
2. Zheng Z, Yang X, Yu Z, Zheng L, Yang Y, Kautz J (2019) Joint discriminative and generative learning for person re-identification. In: IEEE conference on computer vision and pattern recognition (CVPR)
3. Gong K, Liang X, Shen X, Lin L (2017) Look into person: self-supervised structure-sensitive learning and a new benchmark for human parsing. CoRR. Available: http://arxiv.org/abs/1703.05446 [Online]
4. Jetchev N, Bergmann U (2017) The conditional analogy gan: swapping fashion articles on people images
5. Mo S, Cho M, Shin J (2018) Instagan: instance-aware image-to-image translation. CoRR. Available: http://arxiv.org/abs/1812.10889 [Online]
6. Pons-Moll G, Pujades S, Hu S, Black M (2017) Clothcap: seamless 4d clothing capture and retargeting. ACM Trans Graph (Proc SIGGRAPH)
7. Yang S, Amert T, Pan Z, Wang K, Yu L, Berg TL, Lin MC (2016) Detailed garment recovery from a single-view image. CoRR. Available: http://arxiv.org/abs/1608.01250 [Online]
8. Karras T, Laine S, Aila T (2018) A style-based generator architecture for generative adversarial networks. CoRR. Available: http://arxiv.org/abs/1812.04948 [Online]

9. Zhang H, Xu T, Li H, Zhang S, Huang X, Wang X, Metaxas DN (2016) Stackgan: text to photo-realistic image synthesis with stacked generative adversarial networks. CoRR. Available: http://arxiv.org/abs/1612.03242 [Online]
10. Zhu J, Park T, Isola P, Efros AA (2017) Unpaired image-to-image translation using cycle-consistent adversarial networks. CoRR. Available: http://arxiv.org/abs/1703.10593 [Online]
11. Cao Z, Hidalgo Martinez G, Simon T, Wei S, Sheikh YA (2019) Openpose: realtime multi-person 2d pose estimation using part affinity fields. IEEE Trans Patt Anal Machine Intell
12. Goodfellow IJ, Pouget-Abadie J, Mirza M, Xu B, Warde-Farley D, Ozair S, Courville A, Bengio Y (2014) Generative adversarial networks
13. Salimans T, Goodfellow I, Zaremba W, Cheung V, Radford A, Chen X (2016) Improved techniques for training gans
14. Heusel M, Ramsauer H, Unterthiner T, Nessler B, Hochreiter S (2018) Gans trained by a two time-scale update rule converge to a local nash equilibrium
15. Wang Z, Bovik A, Sheikh H, Simoncelli E (2004) Image quality assessment: from error visibility to structural similarity. IEEE Trans Image Process 13(4):600–612
16. Sanodiya RK, Mathew J, Saha S, Thalakottur MD (2019) A new transfer learning algorithm in semi-supervised setting. IEEE Access 7:42 956–42 967
17. Sanodiya RK, Yao L (2020) Unsupervised transfer learning via relative distance comparisons. IEEE Access 8:110 290–110 305
18. Sanodiya RK, Mathew A, Mathew J, Khushi M (2020) Statistical and geometrical alignment using metric learning in domain adaptation. In: 2020 International joint conference on neural networks (IJCNN). IEEE, pp 1–8

Attention Mechanism in Convolutional Recurrent Neural Network for Improving Recognition Accuracy in Printed Devanagari Text

Shaheera Saba Mohd Naseem Akhter and Priti P. Rege

1 Introduction

Devanagari text image recognition is the method that aims to automatically recognize the Devanagari text from the document images into machine-encoded text. Devanagari optical character recognition (OCR) has a broad research area due to its extensive applications in recognizing the text in scene images [1, 2], vehicle number plate recognition [3], digitizing the ancient literature [4] to preserve the cultural heritage of India, etc. Devanagari script belongs to the Brahmic family of scripts. The people from the southeast part of Asia, like India, Tibet, and Nepal, speak various languages that are written using Devanagari. About 600 million global speakers use the Devanagari script in 120 languages. Hindi, Nepali, and Sanskrit are among them. Most of the ancient literature is written in Sanskrit and Hindi. Despite the countless demands, the accuracy of the Devanagari OCR is not at par with that of Latin. Problems primarily come from several word-formation complexities in the Devanagari script and the lack of an open dataset required to train the OCR system.

Devanagari's script covers 14 vowels and 33 consonants [5]. Figure 1 shows the character set of the script. The script has a horizontal line called *"Shirorekha"* on every character and word (see Fig. 2). The *modifier* is a modified character in Devanagari formed using a consonant followed by a vowel.

In some cases, the consonant is followed by one or more consonant(s) to form a *compound* character, also called *conjuncts*, and has an orthographic shape [6]. The large number of distinct characters in Devanagari is due to many *modifiers* and *compound* characters, which makes the Devanagari OCR challenging.

S. S. M. N. Akhter (✉) · P. P. Rege
College of Engineering, Pune, Maharashtra, India
e-mail: sna18.extc@coep.ac.in

P. P. Rege
e-mail: ppr.extc@coep.ac.in

© The Author(s), under exclusive license to Springer Nature Singapore Pte Ltd. 2022   141
J. Mathew et al. (eds.), *Responsible Data Science*, Lecture Notes
in Electrical Engineering 940, https://doi.org/10.1007/978-981-19-4453-6_10

Fig. 1 Devanagari character
set

Vowels: अ आ इ ई उ ऊ ऋ ए ऐ ओ औ
Modifiers: पु पू पे पै पो पौ पृ पी पा पि
Consonants: क ख ग घ ङ च छ ज झ ञ ट ...
Conjuncts: क्र क्ल क्स क्ष ख्म ख्न न्स प्र फ्र र्त र्च श्र ...

Fig. 2 Example of
Devanagari word

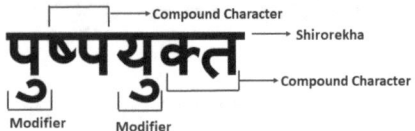

Recently, many improvements have been made in the Devanagari text OCR with various deep learning models [7–9]. The improvements are possible owing to the presence of better deep learning models (like convolutional neural network (CNN), recurrent neural network (RNN), etc.), better regularizers and optimization methods, and availability of better GPUs. Researchers have extensively used convolutional recurrent neural network (CRNN) as one of the best solutions for Devanagari OCR [9]. The current best solutions for Devanagari text recognition are insufficient to resolve the above challenges. Encouraged by recent advances in deep learning, we propose an attention mechanism-based encoder–decoder model for increasing the Devanagari text recognition accuracy. Moreover, the attention mechanism has gained better performance on many text recognition tasks for various other languages [10, 11]. The encoder of the proposed network is a CNN responsible for extracting the high-level features of the input image and arranging them in a proper sequence. The feature sequence from the encoder is given as input to the decoder. The decoder has a BLSTM network with an attention mechanism [12] and a connectionist temporal classification (CTC) [13] layer. The attention mechanism in the decoder provides attention to the relevant features of the encoder required to predict the text from the input image. The entire network is trained end to end, without applying any pre- or post-processing to the input images. Hindi is one of the languages of India and is written using the Devanagari script. In this work, we have constructed the dataset of Hindi word images for training the proposed system. The details of the proposed model are shown in Sect. 3.

The significant contributions of this work are summarized as follows:

- We have proposed an attention-based CRNN model to improve the recognition accuracy of Devanagari text images for the Hindi language.
- As the deep learning model has to learn millions of parameters, the model may not perform well if the number of sample images in the dataset is significantly small. Due to the smaller size of standard dataset from IIIT-H [1], we have curated two datasets, viz. a synthetic dataset and a real dataset. The synthetic dataset has 10000 images, and the real dataset consists of 8000 images of words taken from headlines of Hindi newspapers.

- We have compared the performance of the proposed model with the current state-of-the-art models used for Devanagari text recognition using the dataset presented during the study.

The paper is structured in different sections. Section 2 gives a brief description of the related work in the area of Devanagari text recognition. Section 3 elaborates the methodology proposed in this study. Section 4 has an experimental setup that explains the dataset and parameter selections for training the network. Section 4 also presents the results and discussions. Section 5 has a conclusion.

2 Related Work

The research in Devanagari OCRs started with the implementation of a traditional image processing system, like template matching and feature-based methods, which can classify the printed Devanagari characters and numerals [14–16]. The algorithms were strictly language and script-dependent. Several feature-based algorithms, like artificial neural network (ANN), hidden Markov model (HMM), support vector machine (SVM), etc., were then used by the researchers to increase the accuracy of Devanagari character classification. Kompalli et al. [17] discussed several challenges of the Devanagari script and the classification of Devanagari characters with the ANN method. Shaw et al. [18] had used chain code and HMM algorithms for Devanagari word recognition. The traditional classifiers used for the Devanagari character recognition could classify only a single character or syllable. At the same time, the traditional methods for text recognition of the whole page required segmentation of the page into line, word, and characters [19]. Different algorithms are available in the literature for Devanagari character segmentation [20]. Due to the complexities of the Devanagari characters, the segmentation algorithm had not proven accurate and introduce recognition errors. Later, with the advent of machine learning methods, researchers had also tried various machine learning algorithms [like CNN, multilayer perceptron (MLP)] to classify the Devanagari isolated characters [21, 22]. Again the machine learning algorithms require the segmentation of Devanagari words to an isolated character. The segmentation of words to characters propagates the error to the recognition stage. Moreover, the main drawback of the traditional and machine learning methods is that they require the segmentation of words to characters as well they are not able to remember the context.

Further, the recognition rates in text for different languages increase with the success of deep learning networks [23–25]. The deep Neural Network (DNN) models can remember the context required for any text recognition system. The RNNs, especially long short-term memory (LSTM) [26], BLSTM [27], and multi-dimensional LSTM [28], are the models which are finely used to remember the context in many text recognition task. A hybrid network created using CNN and RNN, i.e., CRNN with CTC loss has proven successful in Devanagari text recognition and given better results than the other methods [9]. Still, the recognition accuracy has not improved as expected. Therefore, we attempted to add an attention mechanism in the existing CRNN model used for Devanagri text recognition for the Hindi language. Due to the Devanagari word-formation complexities, the attention mechanism will focus on the relevant features to give a better result. The primary role of the attention mechanism is to provide more attention to the critical features from the decoder required to predict the text.

3 Methodology

This section describes the proposed attention-based CRNN model used for Devanagari text recognition for the Hindi language. The proposed end to end network has an encoder, an attention network, and a decoder, as shown in Fig. 3. We first illustrate the structure of the encoder and a CNN followed by the explanation of the decoder network (BLSTM network with attention mechanism).

3.1 Encoder: Deep CNN

The encoder of the proposed network is a deep CNN with a series of convolutional and max-pooling layers (refer to encoder part of Fig. 3). There are seven convolution and four max-pool operations performed in the encoder. A filter bank is applied for every convolution operation. Each pixel of the feature maps obtained after all convolution operations is passed to the nonlinearity using rectified linear unit (ReLU). We have also applied batch normalization (BN) [29] to the output obtained after the nonlinearity. Some of the output feature maps are max-pooled to reduce the number of the parameters of the network. We have fixed the size of the input image to 32×128. Table 1 gives the deep CNN configuration of the encoder.

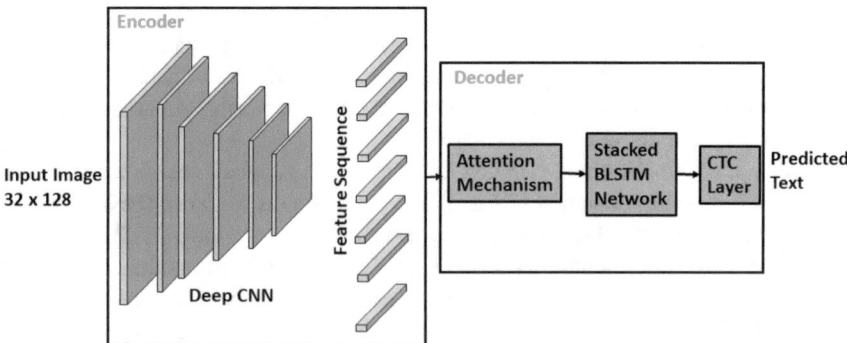

Fig. 3 Block Diagram of the proposed network

Table 1 Deep CNN configuration in the encoder

Blocks	Operations		
	Input Layer		
Block1	(32 × 128 × 1)		
Block2	Conv1 layer [3 × 3] 32 × 128 × 64	Max-pool1 layer 16 × 64 × 64	[2 × 2]
Block3	Conv2 layer [3 × 3] 16 × 64 × 128	Max-pool2 layer 8 × 32 × 128	[2 × 2]
Block4	Conv3 layer [3 × 3] 8 × 32 × 256		
Block5	Conv4 layer [3 × 3] 8 × 32 × 256	Max-pool3 layer 4 × 32 × 256	[1 × 2]
Block6	Conv5 layer[3 × 3] 4 × 32 × 512		
Block7	Conv6 layer [2 × 2] 4 × 32 × 512	Max-pool4 layer 2 × 32 × 512	[1 × 2]
Block8	Conv7 1 × 31 × 512		
Block9	Feature Sequence of 31x512 is generated		

The filter size of every operation is shown in a square bracket. The number in red color is the number of filter operations performed for the layers

The output feature maps found after the seventh convolutional layer are arranged in a feature sequence (i.e., $x = x_1, x_2, x_3, \ldots x_T$) so that it can be adequately fed to the decoder of the system (see Fig. 3). The sequence of features has a high level of semantic information due to CNN's powerful feature extraction ability. Moreover, every feature sequence, i.e., $x_j s$ has a local image region in the input image called receptive fields, which is possible due to the translation invariance property of CNN.

3.2 Decoder: Attention mechanism with BLSTM network

The decoder of the proposed network consists of an attention mechanism and a BLSTM network. The feature sequence of the input images obtained from the encoder is not equally important to the text prediction. Therefore, it is essential to give more attention to the more responsible feature for predicting the text from the input images. Thus, we have implemented an attention mechanism [12] between the encoder and the BLSTM network. The more critical feature sequences are then passed to the decoder to predict the system's respective labels. We have used a weighted attention mechanism, which creates the weighted average of the feature sequences and generates the context vector. The context vector is given as input to the BLSTM network. The attention mechanism structure is shown in Fig. 4.

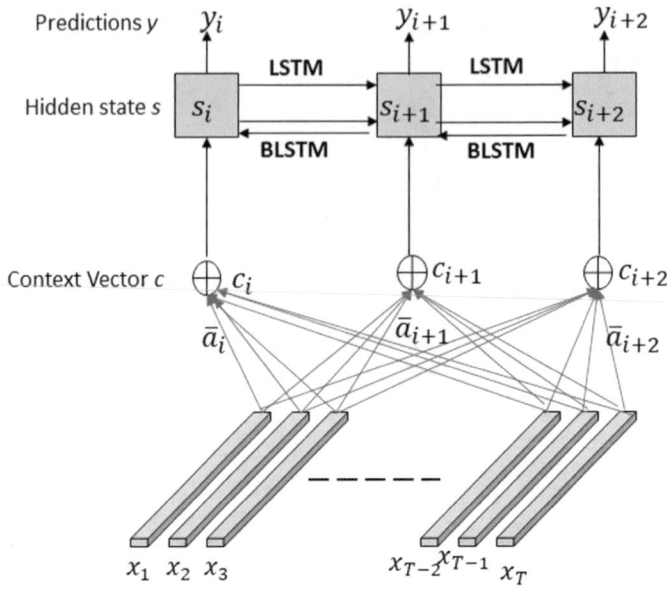

Fig. 4 Attention mechanism placed between encoder and decoder

The context vector is calculated by Eq. 1.

$$c_i = \sum_{j=1}^{T_x} a_{ij} x_j \tag{1}$$

where c_i is the context vector and x_j represents feature sequences of the encoder of the system. aij denotes the weights assigned for the feature sequence x_j, and T_x is the number of feature sequences. The weight vector, a_{ij}, is calculated as (Eq. 2),

$$a_{ij} = f(s_{i-1}, x_j) \tag{2}$$

Function f correlates the encoder to the decoder and is a three-layered feed-forward neural network (FFNN). The equations of the FFNN are given by Eq. 3, 4, and 5.

$$h_{ij} = \tanh(w_{11} * x_j + w_{12} * s_{i-1} + b) \tag{3}$$

$$e_{ij} = h_{ij} * w_{21} \tag{4}$$

Fig. 5 Bidirectional LSTM

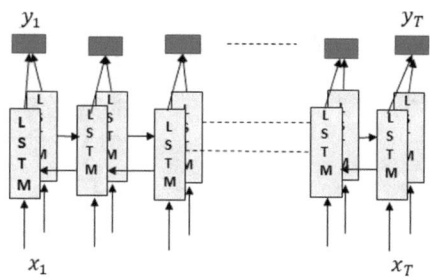

$$a_{ij} = \frac{e^{(e_{ij})}}{\sum_{k=1}^{T_x} e^{(e_{ij})}} \tag{5}$$

where, h_{ij} is the output vector of the first layer of FFNN for the feature sequence x_j and s_{i-1}. w_{11} and w_{12} are the weights of the first layer. w_{21} and b are the weight and bias for the second layer of the neural network. e_{ij} signifies the score of the second layer of the FFNN. We apply a softmax operation to e_{ij} to get the vector a_{ij}. Using vector a_{ij}, we calculate the context vector c_{ij}. c_{ij} is then given as an input to the BLSTM network and subsequently fed to the CTC layer to predict the output. The hyperparameter of the attention mechanism is selected by trial-and-error selection. The selected hyperparameter of the attention mechanism gives the best result.

For the text recognition task, the information from both the direction of the feature sequence is essential. Therefore, two LSTMs are combined in forwarding and backward directions to form bidirectional LSTM (BLSTM) (as shown in Fig. 5). In this study, we have used a stacked BLSTM network. The BLSTM is placed after the attention network. We train the proposed system with the three datasets and evaluate CTC loss results. The number of hidden neurons in the BLSTM is 128, with the output softmax layer with units equal to the total number of Unicode characters set in Hindi.

4 Experiments

4.1 Dataset

In this work, we have trained and tested the proposed network on the three datasets, viz. a synthetic dataset, a real dataset, and the scene images dataset from IIIT-H [1]. The synthetic dataset comprises 10,000 images created using Microsoft Paint. Each image has a Hindi word written. The images from the synthetic dataset are manually corrupted with different types of noises to generalize the results of the system. The

Fig. 6 **a** Sample images
from the synthetic dataset,
and **b** sample images from
the real dataset

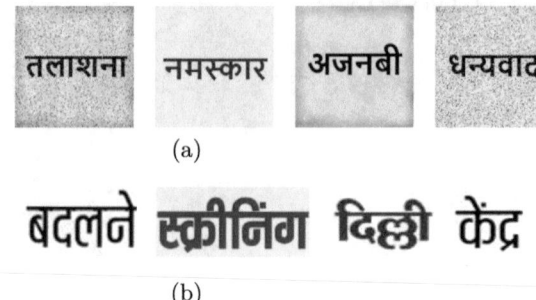

(a)

(b)

real-dataset images are collected from the headlines of the Hindi newspaper articles. We extracted a total of 8800 words from the headlines. using the You Only Look Once [YOLO] object detector [30], treating the words as an object. To compare the performance of our proposed model, we considered the publicly available dataset with scene images provided by IIIT-H. The dataset has 1740 Hindi scene images. Figure 6 shows the sample datasets from the synthetic and real datasets.

The images from the three datasets are augmented using different augmentation methods (like rotation, scaling, blur, etc.) to increase the sample count in every dataset. Also, the images from each dataset are annotated using the Unicode labeling method.

4.2 Implementation Details

The network is trained on NVIDIA Tesla V100-DGXS-32GB with 32 GB of GPU memory using Keras. We choose a batch size of 256. The proposed system generalizes well using 100 epochs. An Adam optimizer [31] with first and second moment exponential decay of 0.5 and 0.999, respectively, is used. The gradients in the encoder are updated using the back-propagation algorithm, while the back-propagation through time (BPTT) [32] is used in the decoder. A dropout of 0.2 is selected during training the model. The kernel and bias of the proposed network are regularized using L2 and L1 regularizers.

4.3 Ablation Study

In this section, we analytically examine the performance of the proposed attention-based CRNN network. We also cite the model in [33] as a baseline to compare the recognition results with the proposed model. The baseline model is the network without attention mechanism. We have trained the proposed and the baseline network

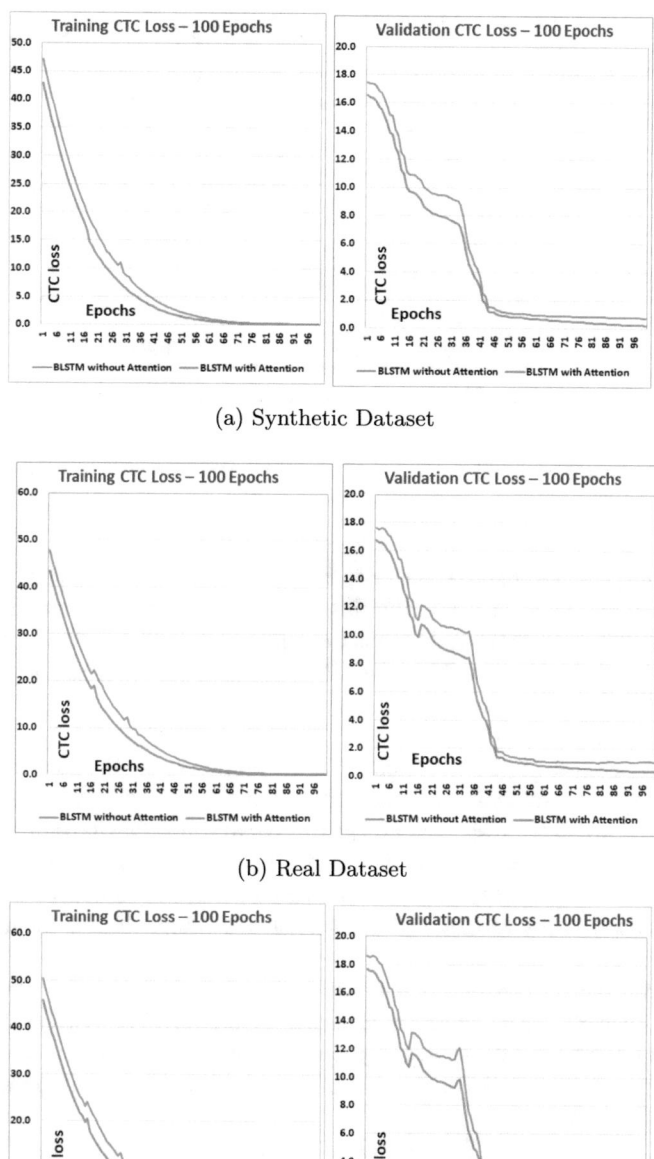

(a) Synthetic Dataset

(b) Real Dataset

(c) Scene image dataset from IIITH

Fig. 7 CTC loss curve during training and validation of the proposed network with baseline model.

Table 2 WER and CER for the proposed network on the three datasets

Dataset type	WER (%)		CER (%)	
	Baseline model	Proposed model	Baseline model	Proposed model
Synthetic dataset	2.12	1.08	0.59	0.05
Real dataset	2.27	1.5	0.63	0.5
IIIT-H dataset	2.43	1.8	0.67	0.23

Table 3 CTC results during testing on three datasets

Dataset type	CTC loss	
	Baseline model	Proposed model
Synthetic dataset	1.32	0.021
Real dataset	1.08	0.035
IIIT-H dataset	1.96	0.061

with the three presented datasets (i.e., trained the three datasets on the model with and without attention mechanism) during this study to simplify the comparisons well. We have individually compared the performance of the system proposed with the baseline network for three datasets with Hindi text images. For comparison purposes, we use the CTC loss [13], word error rate [WER (%)], and character error rate [CER (%)] as evaluation metrics.

Figure 7a–c show the CTC loss curve of the proposed and the baseline network during training and validation of the model on the synthetic dataset, real dataset, and scene image dataset, respectively. The loss curves show that the proposed model generalizes well on three datasets. It has also been noted that the best results are achieved when the proposed model is trained on the synthetic dataset as the other two datasets have many variations in fonts, style of writing, affected by more noise, etc.

We analyze the performance of the proposed model quantitatively by testing one-by-one the images from the three test datasets, respectively. The WER (%) and CER (%) results for the three datasets are given in Table 2. We found that the proposed model has a reduction in WER (%) and CER (%) values compared to the baseline network trained on the three datasets.

We have also listed the CTC loss values in Table 3 while testing the images from the three datasets. The CTC loss value lessens for the proposed model compared to the baseline network.

Table 4 shows the qualitative results when the images are tested on the baseline and proposed model. It has been noted that the text detected using the baseline model has many corrections needed, while the text detected using our proposed model has less number of the error to be corrected and are independent of variations in font size. The incorrect texts detected are highlighted in red color in Table 4. There are some observations we have made during the testing of results. We found that the

Table 4 Qualitative results when the images are tested on the baseline and proposed model

Original image	Ground truth	Detected text with the baseline model	Detected text with the proposed model
जुम्बा	जुम्बा	जुम्वा	जुम्बा
सफर	सफर	अफर	सफर
बैठेंगे	बैठेंगे	बैढेंगे	बैठेंगे
अक्तूबर	अक्तूबर	अवतूबर	अवतूबर
तलाशना	तलाशना	तलासना	तलाशना
स्क्रीनिंग	सक्रीनिंग	सक्रीनिंग	सक्रीनिंग

proposed model sees confusing alphabets, 'स,''ठ,' 'ब,'etc., correctly compared to the baseline model. The conjunct like 'क्त'is detected wrongly by both the models. The conjuncts like 'म्बा'and 'क्री'are seen correctly by the proposed model.

5　Conclusion

We have proposed an attention CRNN network, an encoder–decoder model, for printed Devanagari text recognition using Hindi as a representative of Devanagari script. The attention-based CRNN has proven to give better results. The experiments conducted using two new datasets to examine the performance of the proposed model show that the proposed network meaningfully improved the recognition performance. The whole network with the attention network can be trained end to end.

References

1. Minesh Mathew MJ, Jawahar CV (2017) Benchmarking scene text recognition in Devanagari, Telugu and Malayalam. In: 2017 14th IAPR international conference on document analysis and recognition (ICDAR). IEEE, pp 42–46
2. Shiravale S, Jayadevan R, Sannakki SS (2021) Recognition of Devanagari scene text using autoencoder CNN. Electron Lett Comput Vis Image Anal 20(1):55–69

3. Sivanandan S, Dhanait A, Saiyyad Y (2012) Automatic vehicle identification using license plate recognition for Indian vehicles. Int J Comput Appl
4. Shah KR, Badgujar DD (2013) Devnagari handwritten character recognition (DHCR) for ancient documents: a review. In: 2013 IEEE Conference on information & communication technologies. IEEE, pp 656–660
5. Vaughan D. The world's 5 most commonly used writing systems. [Online]. Available: https://www.britannica.com/list/the-worlds-5-most-commonly-used-writing-systems
6. Pal U, Chaudhuri BB (2004) Indian script character recognition: a survey. Pattern Recogn
7. Kartik Dutta MM, Krishnan P, Jawahar CV (2017) Towards accurate handwritten word recognition for Hindi and Bangla. In: National conference on computer vision, pattern recognition, image processing, and graphics. Springer, Singapore, pp 470–480
8. Dutta K (2019) Handwritten word recognition for Indic & Latin scripts using deep CNN-RNN hybrid networks. Ph.D. dissertation, International Institute of Information Technology, Hyderabad
9. Mehrotra K, Gupta MK, Khajuria K (2019) Collaborative deep neural network for printed text recognition of Indian languages. In: 2019 Fifth international conference on image information processing (ICIIP). IEEE
10. Sheng F, Zhai C, Chen Z, Xu B (2017) End-to-end Chinese image text recognition with attention model. In: International conference on neural information processing. Springer, Cham, pp 180–189
11. He H, Li J (2019) Attention-based deep neural network and its application to scene text recognition. In: 2019 IEEE 11th International conference on communication software and networks (ICCSN). IEEE, pp. 672–677
12. Lee C-Y, Osindero S (2016) Recursive recurrent nets with attention modeling for OCR in the wild. In: Proceedings of the IEEE conference on computer vision and pattern recognition
13. Graves A, Fernández S, Gomez F, Schmidhuber J (2006) Connectionist temporal classification: labelling unsegmented sequence data with recurrent neural networks. In: Proceedings of the 23rd International conference on machine learning (ICML), pp 369–376
14. Palit S, Chaudhuri BB (1995) A feature-based scheme for the machine recognition of printed devanagari script. In: Das PP, Chatterjee BN (eds) Pattern recognition, image processing and computer vision, pp 163–168
15. Pal U, Chaudhuri BB (2005) Printed Devanagari script OCR system. Vivek-Bombay 10:12–24
16. Pal U (2004) Indian script character recognition: a survey. Pattern Recogn 37:1887–1899
17. Kompalli S, Nayak S, Govindaraju V (2005) Challenges in OCR of Devanagari documents. In: Eighth international conference on document analysis and recognition (ICDAR'05). IEEE, pp 327–331
18. Shaw B, Parui SK, Shridhar M (2008) Offline handwritten Devanagari word recognition: a holistic approach based on directional chain code feature and HMM. In: 2008 International conference on information technology. IEEE, pp 203–208
19. Bansal V, Sinha MK (2001) A complete OCR for printed Hindi text in Devanagari script. In: Proceedings of sixth international conference on document analysis and recognition. IEEE Computer Society, pp 203–208
20. Garain U, Chaudhuri BB (2002) Segmentation of touching characters in printed Devnagari and Bangla scripts using fuzzy multifactorial analysis. IEEE Trans Syst Man Cybern Part C (Appl Revi) 32(4)
21. Singh R, Yadav CS, Verma P, Yadav V (2010) Optical character recognition (OCR) for printed Devnagari script using artificial neural network. Int J Computer Sci Commun 1(1):91–95
22. Deore SP, Pravin A (2020) Devanagari handwritten character recognition using fine-tuned deep convolutional neural network on trivial dataset. Sadhana 45(1):1–13
23. He K, Zhang X, Ren S, Sun J (2016) Deep residual learning for image recognition. In: Proceedings of the IEEE conference on computer vision and pattern recognition (CVPR), pp 770–778
24. Szegedy C, Ioffe S, Vanhoucke V, Alemi AA (2017) Inception-v4, inception-resnet and the impact of residual connections on learning. In: Proceedings of the AAAI conference on artificial Intelligence

25. Krizhevsky A, Sutskever I, Hinton GE (2012) ImageNet classification with deep convolutional neural networks. In: Advances in neural information processing systems (NIPS), pp 1097–1105
26. Carbune V, Gonnet P, Deselaers T, Rowlwy HA, Daryin A, Calvo M, Wang L-L, Keysers D, Feuz S, Gervais P (2020) Fast multi-language lstm-based online handwriting recognition. Int J Doc Anal Recogn (IJDAR) 23(2):89–102
27. Sankaran N, Jawahar CV (2012) Recognition of printed Devanagari text using BLSTM neural network. In: Proceedings of the 21st International conference on pattern recognition (ICPR2012), Tsukuba, Japan, pp 322–325
28. Chavan V, Malage A, Gupta MK (2017) Printed text recognition using BLSTM and MDLSTM for Indian languages. In: Fourth international conference on image information processing (ICIIP), Shimla, India, pp 1–6
29. Ioffe S, Szegedy C (2015) Batch normalization: accelerating deep network training by reducing internal covariate shift. In: International conference on machine learning. PMLR, pp 448–456
30. Redmon J, Divvala S, Girshick R, Farhadi A (2016) You only look once: unified, real-time object detection. In: Proceedings of the IEEE conference on computer vision and pattern recognition
31. Kingma DP, Ba J (2014) Adam: a method for stochastic optimization. *arxiv:1412.6980.* [Online]. Available: https://arxiv.org/abs/1412.6980
32. Werbos PJ (1990) Backpropagation through time: what it does and how to do it. In: Proceedings of the IEEE
33. Shi B, Bai X, Yao C (2016) An end-to-end trainable neural network for image-based sequence recognition and its application to scene text recognition. IEEE Trans. Pattern Anal. Mach. Intell. 39(11):2298–2304

Joint Learning for Multitasking Models

Ajai John Chemmanam⑩ and **Bijoy A. Jose**⑩

1 Introduction

Artificial intelligence using neural networks has made tremendous progress in different areas of computer vision such as image classification, object detection, image segmentation and keypoint estimation. State-of-the-art models are tuned to perform well in individual tasks on a specific dataset and often fails to achieve comparable performance when given another related task. Traditional neural network model development typically cares about optimising a certain key parameter such as benchmark scores on a curated dataset for a specific task. Real-life applications of AI models often require multiple models to be used sequentially or as an ensemble to address different business use cases. The pre- and post-processing steps for each model will be an overhead, especially when used sequentially. Also the computations done by initial models are mostly discarded as the subsequent models try to extract the features again.

Multitask learning aims to perform multiple related tasks at the same time. The system learns to do the different related tasks simultaneously by learning their correlation for improved performance. From a biological aspect, we all learn new things based on our prior knowledge on related tasks. For example, we are able to learn new languages by understanding the meaning of words and comparing their usage in another known language. The multitasking learning approach not only saves time to train multiple models on different tasks, but also be able to save time and com-

A. J. Chemmanam
Cyber Physical Systems Lab, Department of Electronics, Cochin University of Science and Technology, Kochi, Kerala, India
e-mail: ajaichemmanam@cusat.ac.in

B. A. Jose (✉)
Cyber Physical Systems Lab, Department of Computer Science, Cochin University of Science and Technology, Kochi, Kerala, India
e-mail: bijoyjose@cusat.ac.in

© The Author(s), under exclusive license to Springer Nature Singapore Pte Ltd. 2022
J. Mathew et al. (eds.), *Responsible Data Science*, Lecture Notes
in Electrical Engineering 940, https://doi.org/10.1007/978-981-19-4453-6_11

putations while doing the inference due to reduced pre- and post-processing. This approach is better suited for mobile devices where the resources are limited.

Our industry partner Vuelogix Technologies, who works in the video analytics and security surveillance domain, required multiple objects to be detected from the video frames, identify the keypoints of people for classifying their body posture and segment individual detected objects for tracking them. Initial use of state-of-the-art models in these tasks quickly became a burden, as multiple models had to be used sequentially or simultaneously. Usage of multiple models along with their pre- and post-processing was computationally heavy and was challenging to enable real-time video analytics. Through this work, we explore the possibility of joint learning approach for multitasking models and build a single common model that can perform different related tasks at once, reducing the computational requirement while maintaining sufficient accuracy levels.

2   Related Works

Multitask learning is a subarea of machine learning and artificial intelligence in which multiple related tasks are solved together. It has been used previously under different names such as joint learning, learning to learn and learning with auxiliary tasks. As part of the literature survey, we explored the previous works in natural language processing (NLP), computer vision, reinforcement learning domains and the commonly used multitask architectures in general.

2.1   *Natural Language Processing*

The concept of multitask learning has been explored more by researchers in the natural language processing (NLP) than in computer vision. Søgaard and Goldberg [21] suggested that some auxiliary tasks require a low level supervision in cases like NLP pre-processing such as parts of speech tagging and entity recognition. Hashimoto et al. [7] predefined a hierarchy for multiple NLP tasks. Liu et al. [12] tried representation learning using multitask network for semantic classification of web query and information retrieval for web search using the same query. Taskonomy [27] developed a visual space of multiple tasks by computing the transfer learning dependencies for over 20 different tasks in 2D and 3D space. Natural language Decathlon [13] is another multitasking challenge in the natural language domain to perform ten different tasks such as question-answering, machine translation, summarization and sentiment analysis.

2.2   Reinforcement Learning

In reinforcement learning domain, DeepMind proposed Distral [23] where they utilised a distilled policy that captures the common behaviour across all tasks. The workers are trained to perform their tasks without sharing the network parameters but limits them to the shared policy. Ramsundar et al. [18] have used multitasking networks for drug discovery. They claim that their multitasking model was able to obtain higher predictive accuracy than single-task methods, and their empirical studies showed that the multitask networks improves as additional tasks and data are added.

2.3   Computer Vision

In the field of computer vision, Kendall et al. [9] used the idea of using different weights for different tasks while calculating the losses. They used an encoder along with multiple decoders to estimate the depth, generate instance and semantic segmentation. They derived the multitask loss function by maximising the Gaussian likelihood considering the uncertainty of each task. Zhang et al. [28] developed facial landmark detection along with head point estimation and facial attributes as auxiliary tasks. They suggested task wise early stopping to improve learning and reduce overfitting of auxiliary tasks. FT-MTL-NET [6] proposed the concept of feature transfer learning between branches of a multitask model and showed promising results in medical imaging applications. Another use case in medical application where multitasking model was tried is in detecting cancerous mass and classifying it as benign or malignant [1]. They used a You Only Look Once (YOLO) model to detect the cancerous area, and then trained a full-resolution convolution network (FrCN) to segment the detection pixel wise and classify it. Recently, mask single shot detector (SSD) [22] was proposed to enhance the performance of popular SSD model to detect small objects. The model contains a detection branch and a segmentation branch to enhance detection features with contextual information.

2.4   General Multitask Architectures

Ruder [19] in his review paper on multitask learning classifies the multitask network architectures into two, hard parameter sharing and soft parameter sharing model architectures. Caruana [2] is one of the pioneers in developing hard parameter sharing multitask network architectures. As shown in Fig. 1, the hard parameter sharing models will have a set of layers that are shared commonly among the different tasks, to learn a more general representation of the data. Based on the features extracted by the common layers, task specific layers are then used to perform their designated

Fig. 1 Hard parameter
sharing

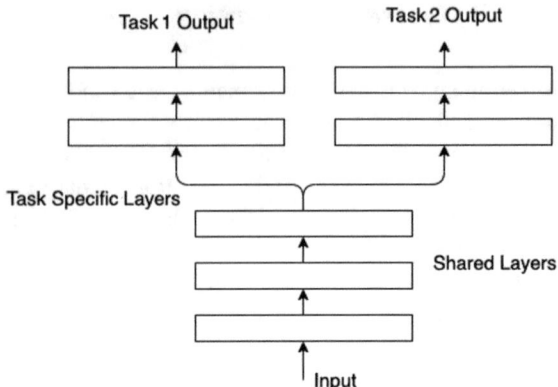

Fig. 2 Soft parameter
sharing

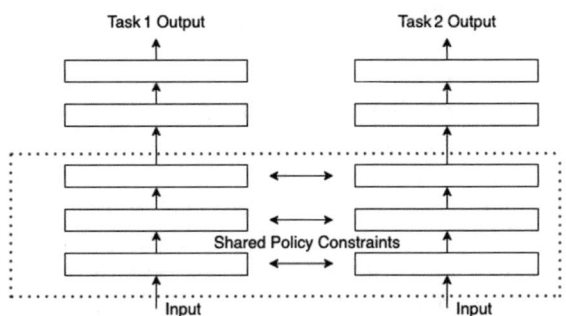

tasks. This approach helps to reduce the chances of overfitting on the data as the shared layers are able to identify a better representation than used for individual tasks.

Soft parameter sharing, as shown in Fig. 2, contains a set of similar non-shared layers for each task. They are like similar independent models performing different tasks in parallel but their parameters are restrained to certain common policies. Certain cases [5] used l2 regularisation and some [26] used trace norm for constraining the shared parameters to be more similar.

Misra et al. [14] use trainable intermediate layers to learn the linear combination of output of previous layers of the parallel architectures. These layers called cross stitch units are placed after pooling and fully connected layers. Similarly, Ruder et al. [20] proposed sluice networks which has trainable parameters that learns to control which parts of the network and amount of layers that needs to be shared between the tasks.

In our context, with a resource constrained computing environment, hard parameter sharing will be the better option, as the use of a common backbone network will have significantly less number of parameters than multiple identical feature extractors. It also reduces the computations (MACS/FLOPS) to be done in it.

3 Proposed Model Architecture

CenterNet [29] is one of the most successful object detection architecture in recent years. It is a single stage model that detects and classifies the object using a single forward pass through the model. It treats the objects to be detected as a single point—the centre of the bounding box. The authors use the keypoint estimation technique to find the centre points and regresses to all other object properties, such as size, 3D location and orientation. Variants of CenterNet are also available, which uses different fully connected encoder–decoder backbone networks such as ResNet, deep layer aggregation (DLA) and hourglass with different accuracy and processing capabilities. The accuracy of the models are measured in average precision (AP) at different intersection over union (IoU) thresholds. The authors achieved 28.1% AP at 142 FPS, 37.4% AP at 52 FPS, and 45.1% AP with multi-scale testing at 1.4 FPS for object detection on MS COCO dataset. Models like CornerNet [10] and ExtremeNet [30] also have similar approaches to the object detection problem. While the former identifies the two corner points, the latter identifies the four corner points through keypoint estimation.

Standing on the shoulders of these giants, we used the CenterNet architecture as our baseline model. A high level architecture of our model is shown in Fig. 3. We replaced the backbone network of the baseline model (shared layers) with a harmonic densely connected network (HarDNet) [3]. HarDNet is a simple U-shaped network with Conv3 × 3, ReLU, bilinear interpolation upsampling and sum-to-1 layer normalisation. The inference time of the model on edge devices was improved by optimising DRAM memory traffic using Nvidia profiler and ARM Scale-Sim software. The two variants HarDNet 68 and HarDNet 85 achieved 76.5 and 78.0 Top1 accuracy on ImageNet classification dataset, respectively. This shared backbone layer is responsible for extracting features common to all tasks.

Then, the individual task heads (branches) perform the task specific computations without sharing their weight parameters. Our multitasking model jointly predicts 17 body keypoints in COCO format as well as the centre of the object using heatmaps.

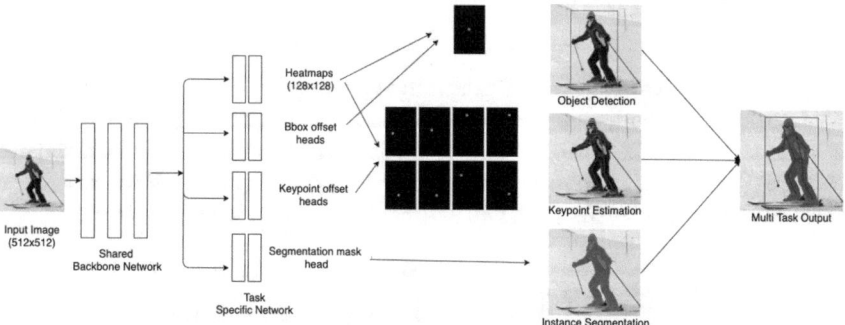

Fig. 3 Multitask architecture

With a stride size of 4, the model predicts heatmaps of size 128×128 for an input image size of 512×512 as shown in Fig. 3. Regression branches are used to predict the height and width of the object. Further, two heads are used to compensate the offset error when upscaling the heatmaps to the original input image size of 512×512 in x and y directions for each of the 18 heatmaps generated.

Segmentation task needs to predict the masks segments for each object instances. Recently, Tian et al. [24] proposed the use of compact dynamically generated instance aware mask heads for instance segmentation. They use a compact fully convolutional layers to generate the instance segmentation without relying on expensive ROI operations. Authors claimed that the different variants of CondInst were able to achieve 35.7 mAP to 40.2 mAP on COCO segmentation tasks. We integrated the CondInst architecture into the segmentation branch. The segmentation mask head in our architecture consists of three 3×3 convolutions with ReLU activation except for the final layer. We chose eight channels for the mask head as suggested in the study [24]. Instead of having fixed filter weights as in traditional convolution layers, the conditional convolutions [25] dynamically generate the weights using a controller network [8]. Total number of weights to be generated $= 2 * (\text{numChannels})^2 + 5 * \text{numChannels} + 1 = 169$. The real magic of multitask learning happens in the shared backbone network which is trained using the multitask loss.

3.1 Multitask Loss

Defining the loss function of the multitasking network is the core idea of joint learning. We can either define the loss for each task and optimise the model separately or optimise them jointly. First approach requires to do alternate training on different task data by calculating the loss and calling the optimisers for each task one after the other. Second approach is more simpler one, compute the sum of loss of individual tasks for each iteration and optimise the model with this joint loss.

As discussed in the previous section, the proposed model predicts 18 keypoints (17 body points as in COCO dataset and the centre of the object). All the ground truth 18 keypoint locations (x, y) are resized to 128×128 output resolution given by (p_x, p_y) and convolved with a Gaussian kernel $\exp^{-\frac{(x-p_x)^2+(y-p_y)^2}{2\sigma^2}}$ where σ is the standard deviation. Following [29], we used a pixelwise logistic regression with focal loss [11] to train the heatmap heads. Focal loss adds a penalty to the loss calculated from unbalanced data, through parameters α and β in Eq. 1. We took $\alpha = 2$ and $\beta = 4$ as seen in CornerNet [10], which uses a similar loss to train their model.

$$L_{\text{focal}} = \frac{-1}{N} \Sigma \begin{cases} (1 - \hat{Y})^\alpha \log \hat{Y} & \text{if } Y = 1 \\ (1 - Y)^\beta (\hat{Y})^\alpha \log (1 - \hat{Y}) & \text{otherwise} \end{cases} \tag{1}$$

Total loss for heatmap and offset heads is given by Eq. (2).

$$L_{\text{heat}} = \Sigma_N(\lambda_{\text{heatmap}} * L_{\text{focal}} + \lambda_{\text{offset}} * L_{\text{offset}}) \tag{2}$$

with $N = 18$ (number of heatmaps). Dice loss [4] is used in the segmentation mask head to calculate the error in mask prediction. Dice loss is given by Eq. (3).

$$L_{\text{dice}} = \frac{2 * \text{Number of overlapping pixels}}{\text{Total number of pixels}} \tag{3}$$

Total multitask loss or joint loss is given by

$$L_{\text{Tot}} = L_{\text{heat}} + \lambda_{\text{reg}} * L_{\text{box}} + \lambda_{\text{segm}} * L_{\text{dice}} \tag{4}$$

While training with the combined loss of different tasks, there are a few considerations to be made. There is a chance for the gradients from different tasks interfering negatively, causing the loss to fluctuate making the learning process unstable. Also there can be cases where one task dominating the others, causing the final loss to be lower but with some underperforming tasks. We try to overcome these issues by weighing the different task losses before combining. Empirical analysis gave the value of λ_{heatmap}, λ_{reg}, λ_{offset} and λ_{segm} as 1, 0.05, 0.5 and 0.1, respectively.

At the same time, adding the different losses will be beneficial in these ways:

- They act as regularisation, smoothing the loss function.
- They give informative priors to the model. It boosts the features that are common to different tasks significantly improving the learning curve.
- Better and faster convergence as the models looks at a more common representation for an optimised solution.
- Self feature attention: If a feature is relevant to more than one task, then it should be more important and better representative of the data. The model will be able to prioritise them over other features.
- Benefits of transfer learning: We already know the benefits of transfer learning, helping the models learn better in another related domain. Jointly learning the different tasks helps the model to learn better and faster utilising these phenomena.
- Avoids overfitting: Losses of auxiliary tasks can help avoid overfitting on any individual tasks or data.

3.2 Training Process

The multitasking model required annotations for detections, keypoints as well as instance segmentations. A dataset with all these different annotations is rare. Most of the existing datasets are curated only for any one single task only. We chose the Common Objects in Context (COCO) dataset, which had annotations for object detection, body keypoints for person, instance, panoptic segmentations etc.

Our train and test setup is done on a system with Intel(R) Core(TM) i9-10900KF CPU @ 3.70 GHz, NVIDIA Geforce RTX 3090 GPU, Pytorch 1.9.0+cu111 and TensorRT 8. We trained a h85 model with HarDNet85 backbone and a lighter h68 model using HarDNet68 backbone. All other parameters related to the architecture of the multitasking model remain the same for both models. The models were trained for 350 epochs with the initial learning rate being 0.01 and a mini-batch of 32 images. Each iteration took 42.5 min on an average with a total of 9 days to train the model. The learning rate is reduced by a factor of 10 at iterations 250 and 300.

During testing and inference, We then use a 3×3 max pooling on the heatmaps to find the peaks on it, eliminating the need of using non-maximum suppression (NMS) [15]. This is because even if multiple objects of the same class are very close together, their centre point will be less likely to be the same. The intensity of the heatmaps gives the confidence for the detection and keypoints.

4   Benchmarks

Our end goal is to use the models trained on custom data for detecting multiple objects from CCTV videos, and classify the body posture of the person using the keypoints identified and extract the instance segmentation masks for removing the background while doing feature extractions in object tracking. However, it is not possible to compare directly the performance of such a model with other models. We evaluated the multitask model performance on COCO dataset-validation split on all the three tasks that the model learned (Tables 1, 2 and 3).

We used the official pycoco tools library to calculate the evaluation metrics. The average precision is a popular metric in object detection. Intersection over union (IoU) is the ratio of area of overlap and area of union between two bounding boxes. The precision and recall are calculated with different IoU thresholds. For example, if IOU threshold =0.5, a detection with 0.7 IOU is considered as true positive (TP) and a detection with 0.3 IOU is considered as a false positive. Precision is calculated as $\frac{TP}{TP+FP}$ and recall $\frac{TP}{TP+FN}$. Average precision is defined as the area under the precision–

Table 1 COCO evaluation BBox results

Model name	AP	AP 50	AP 75	AP small	AP medium	AP large
Ours-h68	0.368	0.582	0.397	0.164	0.556	0.498
Ours-h85	0.412	0.599	0.452	0.131	0.503	0.709

Table 2 COCO evaluation segmentation results

Model name	AP	AP 50	AP 75	AP small	AP medium	AP large
Ours-h68	0.295	0.543	0.297	0.102	0.452	0.437
Ours-h85	0.382	0.643	0.406	0.135	0.524	0.638

Table 3 COCO evaluation keypoint results

Model name	AP	AP 50	AP 75	AP medium	AP large
Ours-h68	0.393	0.643	0.408	0.428	0.418
Ours-h85	0.530	0.786	0.563	0.449	0.664

recall curve. We calculated the average precision over all IOU thresholds (AP), AP at IOU thresholds 0.5 (AP50) and 0.75 (AP75).

The model with HarDNet85 (h85) backbone network gave 43.6 AP on object detection, 38.2 AP on instance segmentation and 53.0 AP on keypoint estimation tasks on evaluation with COCO validation data. Smaller model with HarDNet68 (h68) backbone network gave 36.8 AP on object detection, 29.5 AP on instance segmentation and 39.3 AP on keypoint estimation tasks. The inference of the h85 model took 0.031 s and h68 model took 0.024s on an average. This translates to around 32.26 frames per second (fps) for h85 model and 41.66 fps for h68 model on the GPU with 1895 MB GPU RAM usage and 5–7% GPU utilisation.

4.1 Comparison with State-of-the-Art Models

Due to the lack of similar multitasking architectures that are designed for exact similar tasks, it is difficult to compare with state-of-the-art models. We considered the single-task models, that had similar architectures to our baseline model for comparison. Our model is evaluated without flip/multiscale testing.

Tables 4, 5, 6 compare the performance of different tasks individually with their baseline models. In Table 4, we can see that the bounding box AP for our h85 multi-

Table 4 Bbox evaluation—comparison with baseline

Model name	AP	AP 50	AP 75
ResNet50 + CondInst	0.358	0.540	0.384
HardNet68 + SSD512	0.317	–	–
HardNet85 + SSD512	0.351	–	–
CenterNet − Hourglass	0.403	0.591	0.440
CenterNet − DLA1x	0.363	–	–
CenterNet − DLA2x	0.374	0.551	0.408
CenterNet − ResNet101 (DCN)	0.346	0.530	0.369
CenterNet − ResNet18 (DCN)	0.281	0.449	0.296
ExtremeNet + Hourglass	0.358	–	–
ExtremeNet + Deep layer aggregation (DLA)	0.330	–	–
Ours-h68	0.368	0.582	0.397
Ours-h85	**0.412**	**0.599**	**0.452**

Bold letters signify highest value

Table 5 Segmentation evaluation—comparison with baseline

Model name	AP	AP 50	AP 75	AP small	AP medium	AP large
ResNet50 + CondInst	0.354	0.564	0.376	**0.184**	0.379	0.469
Ours-h68	0.295	0.543	0.297	0.102	0.452	0.437
Ours-h85	**0.382**	**0.643**	**0.406**	0.135	**0.524**	**0.638**

Bold letters signify highest value

Table 6 Keypoint evaluation—comparison with baseline

Model name	AP	AP 50	AP 75	Speed (fps)
Hourglass-104	**64.0**	–	–	6.6
DLA-34	58.9	–	–	23
Ours-h68	0.393	0.643	0.408	**41.66**
Ours-h85	0.530	0.786	0.563	**32.26**

Bold letters signify highest value

tasking model is better than original conditional instance segmentation model with ResNet, SSD model with similar HarDNet backbone networks, variants of Center-Net and ExtremeNet models. The lighter multitasking model also has comparable performance with the best performing CenterNet baseline models. Table 5 compares our models with the baseline conditional instance segmentation model, for the instance segmentation task. The h85 model outperformed the baseline model on AP at different IOUs and for medium and large objects. Table 6 shows that we achieved acceptable performance in keypoint estimation task with less than half the computation time required for other models. Both our models perform more than five times faster than the current state-of-the-art hourglass model in estimating the keypoints.

5   Observation on Real Data

The proposed model architecture was designed keeping the requirements of our industry partner in mind. The multitasking architecture was retrained on custom data which is directly sourced from real-life CCTV footage. We tried to detect person along with their head and identify the helmets or caps they are wearing. The body keypoints estimated by the same model was used for pose classification. Mask generated for each object instance helps to remove the background and effectively track them. Without the auxiliary tasks, the detection of helmet/cap was a challenging task as the number of false predictions was high even in frames without a person. The keypoint estimation helped to detect people better and the model learned that the helmet/cap/head was more common overlapping the person detection. Another observation is that the segmentation task helped to better constrain the detected bounding boxes to the exact edges of the objects rather than predicting very large or smaller bounding boxes than the object. We attribute these improvements as a sort of self-attention focusing on relevant features by the network. However, a more detailed ablation study may be required to confirm these assumptions. As a future work, we are also planning to use these multitasking models in a previously developed robotic testbed [16, 17], and measure the impact of the additional auxiliary tasks in such resource constrained embedded systems.

6 Conclusion

Most of the existing state-of-the-art models are optimised for certain benchmark parameters on curated datasets for specific tasks. From an industrial point of view, business use cases often require results from multiple models, used sequentially or as an ensemble, for generating meaningful business insights. This can be computationally heavy for a real-time analytics pipeline. Also there are significant overheads in pre- and post-processing for each of the models and are sub-optimal when needs to have sequential processing. Each model ends up extracting their own features and are not able to take advantage of pre-computed features.

We developed a multitasking model that is able to do multiple tasks in a single forward pass. The joint learned model is able to do multiple related tasks—object detection, keypoint estimation for body posture analysis and instance segmentation for improved tracking. We achieved 41.2 mAP on object detection task, 38.2 on segmentation task and 53.0 on keypoint estimation task, when evaluated on COCO validation data. The model was able to achieve 32.26 frames per second, while a lighter version achieved 41.66 fps.

The joint learned models were compared against state-of-the-art single-task models that had similar architectures to our baseline. Our models outperformed these models in terms of accuracy in bounding box and segmentation tasks. We achieved comparable performance in keypoint estimation task with nearly 5–7 times faster inference time than others. We attribute the improved performance of the model to the carefully chosen multitasking architecture and the optimised backbone network. Our assumption is that the use of auxiliary tasks enabled the model to become more robust and generalised one, reducing the chances of overfitting and understanding the complex relationship between the tasks. As future work, we are planning to implement the multitasking model in previously developed robotic testbed [16, 17] and confirm these assumptions through an ablation study, measuring the impact of the additional auxiliary tasks.

The use of a multitasking model helped us to avoid using multiple models for different use cases and reduce their pre–post computational cost and time complexity. This enabled us to process data in real time with sufficient accuracy levels.

Acknowledgements This research work was funded and supported by Vuelogix Technologies Pvt Ltd, Confederation of Indian Industry (CII) and Department of Science and Technology (DST), Government of India through Prime Minister's Fellowship for Doctoral Research 2020 (DST-SERB) and DST-ICPS project No. DST/ICPS/CPS-Individual/2018/392 titled "Energy efficient cyber security implementations for Internet of Things".

References

1. Al-Antari MA, Al-Masni MA, Choi MT, Han SM, Kim TS (2018) A fully integrated computer-aided diagnosis system for digital X-ray mammograms via deep learning detection, segmentation, and classification. Int J Med Inf 117:44–54
2. Caruana R (1997) Multitask learning. Mach Learn 28(1):41–75
3. Chao P, Kao CY, Ruan YS, Huang CH, Lin YL (2019) Hardnet: a low memory traffic network. In: Proceedings of the IEEE/CVF international conference on computer vision, pp 3552–3561
4. Deng R, Shen C, Liu S, Wang H, Liu X (2018) Learning to predict crisp boundaries. In: Proceedings of the European conference on computer vision (ECCV), pp 562–578
5. Duong L, Cohn T, Bird S, Cook P (2015) Low resource dependency parsing: cross-lingual parameter sharing in a neural network parser. In: Proceedings of the 53rd annual meeting of the Association for Computational Linguistics and the 7th international joint conference on natural language processing (volume 2: short papers), pp 845–850
6. Gao F, Yoon H, Wu T, Chu X (2020) A feature transfer enabled multi-task deep learning model on medical imaging. Expert Syst Appl 143:112957
7. Hashimoto K, Xiong C, Tsuruoka Y, Socher R (2016) A joint many-task model: growing a neural network for multiple NLP tasks. arXiv preprint arXiv:1611.01587
8. Jia X, De Brabandere B, Tuytelaars T, Gool LV (2016) Dynamic filter networks. Adv Neural Inf Process Syst 29:667–675
9. Kendall A, Gal Y, Cipolla R (2018) Multi-task learning using uncertainty to weigh losses for scene geometry and semantics. In: Proceedings of the IEEE conference on computer vision and pattern recognition, pp 7482–7491
10. Law H, Deng J (2018) Cornernet: detecting objects as paired keypoints. In: Proceedings of the European conference on computer vision (ECCV), pp 734–750
11. Lin TY, Goyal P, Girshick R, He K, Dollár P (2017) Focal loss for dense object detection. In: Proceedings of the IEEE international conference on computer vision, pp 2980–2988
12. Liu X, Gao J, He X, Deng L, Duh K, Wang YY (2015) Representation learning using multi-task deep neural networks for semantic classification and information retrieval
13. McCann B, Keskar NS, Xiong C, Socher R (2018) The natural language decathlon: multitask learning as question answering. arXiv preprint arXiv:1806.08730
14. Misra I, Shrivastava A, Gupta A, Hebert M (2016) Cross-stitch networks for multi-task learning. In: Proceedings of the IEEE conference on computer vision and pattern recognition, pp 3994–4003
15. Neubeck A, Van Gool L (2006) Efficient non-maximum suppression. In: 18th international conference on pattern recognition (ICPR'06), vol 3. IEEE, pp 850–855
16. Nithin P, Francis A, Chemmanam AJ, Jose BA, Mathew J (2019) Face tracking robot testbed for performance assessment of machine learning techniques. In: 2019 7th international conference on smart computing and communications (ICSCC). IEEE, pp 1–5
17. Nithin PB, Chemmanam AJ, Jose BA, Mathew J et al (2020) Interactive robotic testbed for performance assessment of machine learning based computer vision techniques. J Inf Sci Eng 36(5)
18. Ramsundar B, Kearnes S, Riley P, Webster D, Konerding D, Pande V (2015) Massively multitask networks for drug discovery. arXiv preprint arXiv:1502.02072

19. Ruder S (2017) An overview of multi-task learning in deep neural networks. arXiv preprint arXiv:1706.05098
20. Ruder S, Bingel J, Augenstein I, Søgaard A (2017) Sluice networks: learning what to share between loosely related tasks, vol 2. arXiv preprint arXiv:1705.08142
21. Søgaard A, Goldberg Y (2016) Deep multi-task learning with low level tasks supervised at lower layers. In: Proceedings of the 54th annual meeting of the association for computational linguistics (volume 2: short papers), pp 231–235
22. Sun C, Ai Y, Wang S, Zhang W (2021) Mask-guided SSD for small-object detection. Appl Intell 51(6):3311–3322
23. Teh YW, Bapst V, Czarnecki WM, Quan J, Kirkpatrick J, Hadsell R, Heess N, Pascanu R (2017) Distral: robust multitask reinforcement learning. arXiv preprint arXiv:1707.04175
24. Tian Z, Shen C, Chen H (2020) Conditional convolutions for instance segmentation. In: Computer vision–ECCV 2020: 16th European conference. Proceedings, Part I 16. Springer, , Glasgow, UK, 23–28 Aug 2020, pp 282–298
25. Yang B, Bender G, Le QV, Ngiam J (2019) Condconv: conditionally parameterized convolutions for efficient inference. arXiv preprint arXiv:1904.04971
26. Yang Y, Hospedales TM (2016) Trace norm regularised deep multi-task learning. arXiv preprint arXiv:1606.04038
27. Zamir AR, Sax A, Shen WB, Guibas LJ, Malik J, Savarese S (2018) Taskonomy: disentangling task transfer learning. In: IEEE conference on computer vision and pattern recognition (CVPR). IEEE
28. Zhang Z, Luo P, Loy CC, Tang X (2014) Facial landmark detection by deep multi-task learning. In: European conference on computer vision. Springer, pp 94–108
29. Zhou X, Wang D, Krähenbühl P (2019) Objects as points. arXiv preprint arXiv:1904.07850
30. Zhou X, Zhuo J, Krahenbuhl P (2019) Bottom-up object detection by grouping extreme and center points. In: Proceedings of the IEEE/CVF conference on computer vision and pattern recognition, pp 850–859

A CNN Approach for Detecting Red and White Lesions in Retinal Fundus Images

Rajesh Kumar and K. V. Pramod

1 Introduction

Diabetic retinopathy (DR) is a leading cause of blindness caused by damage to the retina from diabetes. It is asymptomatic in early stages but affects 40% of all patients with prolonged diabetes. Early detection through retinal screening is the key to prevention of blindness. A trained retinal expert can diagnose early symptoms of diabetic retinopathy using a retinal imaging device called a *fundus camera*. But, resource constraints and lack of trained manpower make it impossible to implement regular retinal screening programmes to all diabetics, particularly in developing countries (Fig. 1).

A WHO report by Mariotti et al. [1] states that there are 39 million blind people in the world, 6% of which is caused by diabetic retinopathy. According to the International Diabetic Federation [2], of the 415 diabetics in the world, 109 million reside in China followed by 69 million in India. The Chennai Urban Rural Epidemiology (CURES) Eye Study, a population-based study by Deepa et al. [3], concluded the overall prevalence of DR was 20.8% in diabetic subjects in Chennai. All of these studies suggest the urgent need to conduct regular retinal screening of diabetic patients.

The proposed work is to develop a computer-assisted retinopathy detection system based on deep learning which can help in first-level screening of the eye, and then, only the suspected cases of retinopathy need to be reviewed by a human expert

R. Kumar (✉) · K. V. Pramod
Cochin University of Science and Technology, Kochi, India
e-mail: rajeshkumar@cusat.ac.in

K. V. Pramod
e-mail: pramodkv4@gmail.com

R. Kumar
Centre for Development of Advanced Computing, Thiruvananthapuram, India
e-mail: rajeshkumar@cdac.in

© The Author(s), under exclusive license to Springer Nature Singapore Pte Ltd. 2022
J. Mathew et al. (eds.), *Responsible Data Science*, Lecture Notes
in Electrical Engineering 940, https://doi.org/10.1007/978-981-19-4453-6_12

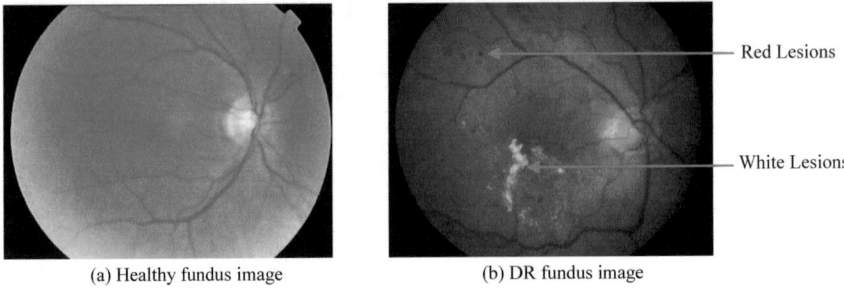

(a) Healthy fundus image   (b) DR fundus image

Fig. 1 **a** Healthy fundus image **b** DR fundus image

for confirmation and further treatment, if required. Attempts to develop such automated methods have been ongoing for over a decade but have not succeeded in achieving performance levels suitable and economically viable for use in large-scale deployment. Scotland et al. [4] did a study on the cost-effectiveness of replacing first-level manual screening with an automated screening software in the national DR screening programme of Scotland. The automated strategy showed a sensitivity of 0.86 compared to manual strategy of 0.87. Thus, the automated grading would give a saving of £201,600 per year to the NHS. Bhaskaranand et al. [5] did an evaluation of EyeArt cloud-based grading system integrating it with EyePACS telemedicine system. The system gave a sensitivity (Sn) of 0.90 and specificity (Sp) of 0.61. EyeArt can be safely integrated into a real-time telemedicine system and achieves a workload reduction up to 62%. Walton et al. [6] did an evaluation of the IRIS automated screening system on 15,015 patients from Harris Health System in Texas. The IRIS system performance was 0.66 Sn and 0.73 Sp. Roy et al. [7] did a pilot study to evaluate the efficacy of an automated retinal grading system in India. The images were analysed by Retmarker software and achieved 0.91 Sn and 0.59 Sp. All these studies have shown the efficacy of automated screening systems whilst bringing economic benefits too.

1.1 Related Works

Machine learning techniques have historically tried to 'mimic' the human learning process aiming to achieve similar results. With traditional machine learning methods, the workflow is almost always *pre-processing->segmentation->feature extraction->classification*. The challenge here is the accurate segmentation of the objects of interest, then extract features of these objects which distinguish them from one-another and across different categories. In this particular problem of automated analysis of diabetic retinopathy from retinal images, this would involve the detection of the lesions like *micro-aneurisms, haemorrhages and hard exudates* which are the key symptoms of the disease. Based on the presence of one or more of these indicators, the retina is classified into different grades like *Non-DR, Mild DR, Moderate DR, Severe NPDR and Proliferative DR*. In a way, traditional approaches bear close resemblance to the diagnosis methods used by the medial experts. Earlier works by Fleming et al. [8] and Goatman [9] proposed morphological tophat transforms and region growing methods to detect micro-aneurisms yielding 0.85 and 0.83 Sn and 0.83 and 0.71 Sp, respectively. Bhalerao et al. [10] propose a method of micro-aneurism detection using LoG followed by matched filtering by circular symmetry operator on DIARETDB1 database with 0.82 Sn and 0.81 Sp. Hatanaka et al. [11] propose a method for detection of haemorrhages and exudates by thresholding on difference image generated by using two different filter masks. They achieved 0.85 Sn in haemorrhages and 0.77 Sn in exudate detection. Niemeijer et al. [12] propose a method for detection of white lesions using a probability-based pixel classification followed by a KNN classifier. They achieved overall 0.95 Sn and 0.88 Sp. Sharath et al. [13–15] used multichannel histogram analysis following by thresholding to detect white lesions achieving 0.88 Sn and 0.95 Sp on DIARECTDB1 data. Further, red lesion detection done using Gaussian matched filter achieved 0.79 Sn. Based on the aggregate of the lesions detected, the overall image level classification measured 0.80 Sn and 0.50 Sp.

The obvious limitation of traditional methods was the reliable and accurate detection of the lesions. A wrongly segmented object will obviously result in wrong feature values being derived from it, likely affecting the performance of the classification method. The second challenge is to extract the 'right features' that have a clinically sound basis and form a mathematical representation for it. At a basic level size, shape, colour/texture, etc. would be features described by the human expert. These could in turn be mathematically represented by area, eccentricity, homogeneity, respectively. As the problem gets complicated in the clinical analysis, so also it gets complicated in the machine learning domain, making them also computationally more intensive and beyond a point, does not yield any better result. The previous works based on traditional image processing and machine learning approaches had achieved accuracy levels of about 80% on large databases although higher results have been reported on smaller databases like DIARETDB1.

The new approaches with deep learning have shown much improved accuracy levels and low computational loads. Typically, using the whole image as an input, use a deep CNN to 'self-learn' against the ground truth marked by the medical expert. The network learns by itself, adjusting the weights. Pratt et al. [16] developed a 10 layer CNN followed by 3 dense layers for DR classification on Kaggle database and achieved a 0.95 Sp and 0.30 Sn. Gulshan et al. [17] used Google's Inception V3 network on Messidor database to get a 0.87 Sn and 0.98 Sp using an ensemble of 10 networks with 22 million parameters. Garcia et al. [18] modified the VGG16 net, by adding a Class Weight in the cost function. They took advantage of the pre-trained VGG16 net producing 0.93 Sp and 0.54 Sn. Shankar et al. [19] propose a Bayesian optimisation of hyperparameters of Inception V4 model and claim accuracy and sensitivity level of over 0.99. Their work is based on MESSIDOR dataset and final classification of the retina. Eftekhari et al. [20] proposed a two-stage CNN method to detect micro-aneurisms using 101×101 pixel image patches and aim to classify each pixel as belonging to micro-aneurism or not. They achieved a sensitivity of 0.769 with a false-positive rate of 8 per image. Zhang et al. [21] propose a DCNN model with multi-layer attention mechanism to detect micro-aneurisms and achieved a sensitivity of 0.868. Both these works were focussed on detection of micro-aneurisms alone and excluded other red lesions and white lesions. Overall, DCNN-based methods tend to produce better results than traditional machine learning methods, but with the drawback that the features being learnt do not necessarily have a mapping to specific symptoms defined in the medical domain. They also require very large datasets to train the network.

2   Materials and Methods

We propose a hybrid approach combining the better aspects of traditional machine learning and deep CNN. Instead of attempting to classify each individual lesion, we group *micro-aneurism* and *haemorrhages* as *red lesions* and *cotton wool spots*, *drussen* and *hard exudates* as *white lesions*. The CNN is trained to detect these clinically significant features, rather than on the final DR grade. The input colour image is split into small patches of 64×64 pixels, and each patch is classified as containing a red lesion, white lesion or no lesion. Using CNN excludes the requirements to segment each lesion, feature extraction and classification of each type. Instead, we focus on detection of the presence of the lesion somewhere within this small patch. Thus, we can train this using much smaller networks, than that would be needed to classify the entire image as DR or non-DR. With this approach, there is clarity as to what each CNN is trying to learn, namely the same clinical symptoms as identified in the medical domain, without having to delineate the boundaries of each in the retinal image.

2.1   Fundus Image and Annotations

For this study, 4260 colour fundus image files were obtained from the Regional Institute of Ophthalmology, Thiruvananthapuram, using Topcon TRC 50DX fundus camera. The images were captured after dilating the pupil, in lossless TIFF format of size 1424×2144 pixels. Of these, 3134 images were annotated and graded by a team of qualified ophthalmologists using a software specifically developed in Matlab for this. The diagnostic quality of the image was assessed as *Good, Average or Poor*. The images were annotated by marking their *x, y* coordinates, and each annotation labelled as one of *Micro-aneurism, Dot Haemorrhage, Flame Haemorrhage, Blot Haemorrhage, Hard Exudates, Drusen and Cotton Wool Spots* that occur in *non-proliferative* stage of DR and other lesions like *Firbro Vascular Proliferation(FVP),*

Table 1 Image annotation details

Annotation type	Count	Annotation type	Count
Micro-aneurism	8475	FVP	579
Dot haemorrhage	3289	IRMA	93
Flame haemorrhage	3603	Laser mark	1598
Blot haemorrhage	3239	Neo-vascularisation	76
Hard exudates	22840	NVD	135
Drussen	1098	NVE	246
Cotton wool spots	1303	Pre-retinal haemorrhage	685
		Venous looping	27
		Vitreous haemorrhage	127

Intraretinal Microvascular Abnormality (IRMA), Neo Vasularisation, Neovasularistion Near Disc (NVD), Neovscularisation Elsewhere (NVE), Pre-retinal haemorrhage, Venous Looping and Vitreous Haemorrhage that occur in proliferative stage. In all 47,413 annotations were marked by the experts. The team also graded the final classification of the retina as *Non-DR, Very Mild, Mild, Moderate, Severe, Very Severe, Non High Risk, High Risk, ADED, Regressed PDR, Other Retinopathy, Insufficient Evidence.* The annotations and grades were stored in '*mat*' file format by the software (Table 1).

2.2 Ethical Considerations

All necessary approvals have been taken from the RIO's Human Ethics Committee for using this image data for academic research vide approval No. 32/HEC/RIOTVPM. Further, a separate approval was obtained vide letter No. CA/218/RIO/16 to publish the results after anonymising the patient details.

2.3 Environment Setup

Matlab script was used to parse the stored annotations and create anonymised dataset for training and validating the CNN. For creating and training the model, CUDA accelerated Tensorflow and Keras libraries were used. The programming language was Python in Jupyter Notebook (Table 2).

Table 2 Environment setup and software tools

Item description	Specification and version
Hardware	Laptop with core i7 CPU 16 GB RAM Nvidia RTX 2070 GPU with 8 GB RAM
Operating system	Windows 10 64 bit
Dataset creation and image pre-processing	Matlab 2021a
Deep learning	Miniconda 3 environment, TensorFlow-GPU ver 1.14, Keras 1.0.8 Cudatoolkit 10.0.130, Python 3.6.7, Jupyter Notebook 1.0 Matlplotlib 3.3.4, Numpy 1.18.1, Scikit-learn 0.24.2

2.4 Training Data and CNN Model

As the annotations were marked and saved using a Matlab script, another Matlab script was written to read and create the image database for training and validation whilst anonymising the patient details. Only, the annotations from images graded as good or average quality by the experts were used in this study, whilst those from images graded as poor were discarded. Since the images were acquired in a controlled environment and poor quality images were discarded, no specific pre-processing steps were applied. A Matlab script was written to parse the annotation files, extract the grade, quality and annotations. Colour image patches of size 64 × 64 pixels were extracted from the respective images surrounding the annotation mark. A random shift is applied in Fig. 2 to the coordinates (x,y) of the annotation to get $(x1,y1)$, then 64×64 patch is extracted with $(x1,y1)$ as the top-left of the patch. This is done to ensure, that the training data do not always have the annotation in the exact centre of the patch.

The above approach was taken for the marked annotations (i.e. having some diagnosis), but we also needed some true negative patches, i.e. which do not have any retinopathy. There are no specific markings in the negative images by the expert, but the image-level grading is available. So, we create some random patches from those images which are graded as *Non-DR* by the expert. About 25 random patches of 64 × 64 each were taken from about 600 non-DR images totalling 15,422 *'no lesion'* patches. In all 56,448 image, patches were generated covering all categories (Table 3). In this study, we are focussed on the early stage DR, and hence, only, micro-anuerism, haemorrhages, cotton wool spots, drussen and hard exudates are considered and grouped into three classes (Fig. 3). Using stratified sampling in, input data were split 80% for training and 20% for testing.

Fig. 2 Creating training patches

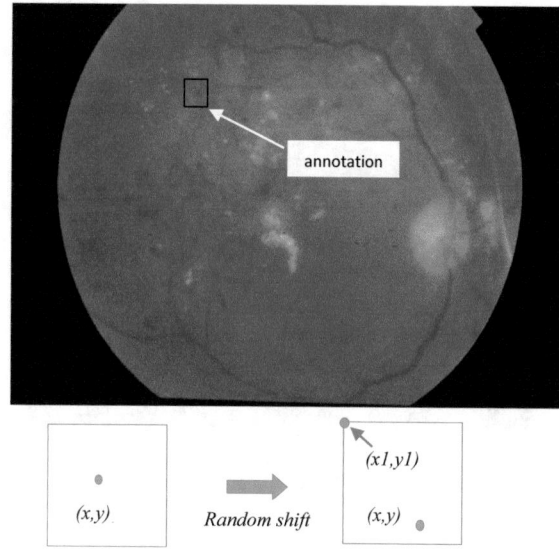

Table 3 Training data grouped into three classes

Annotation	Class	Count
No diagnosis	Class 0 (no lesion)	15422
Micro-aneurism	Class 1 (red lesions)	17789
Dot haemorrhage		
Flame haemorrhage		
Blot haemorrhage		
Hard exudates	Class 2 (white lesions)	23237
Drussen		
Cotton wool spots		

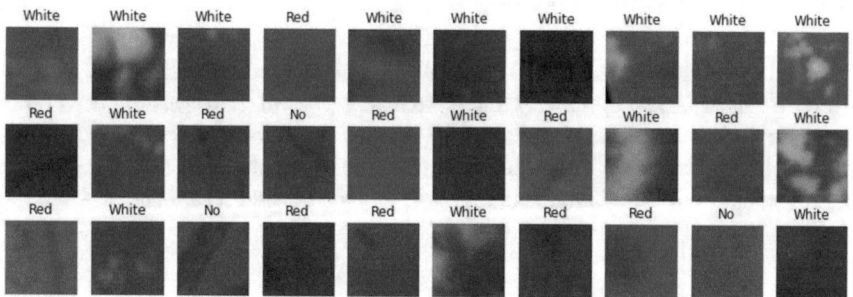

Fig. 3 Sample 64 × 64 patches of red lesion, white lesion and no lesion categories

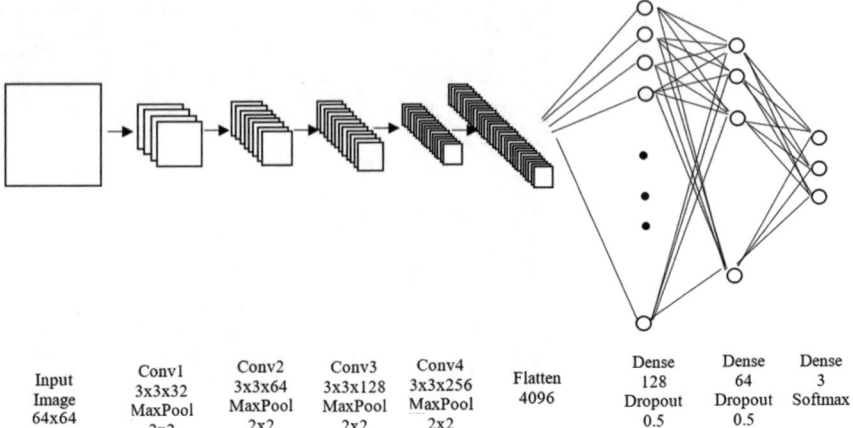

| Input Image 64x64 | Conv1 3x3x32 MaxPool 2x2 | Conv2 3x3x64 MaxPool 2x2 | Conv3 3x3x128 MaxPool 2x2 | Conv4 3x3x256 MaxPool 2x2 | Flatten 4096 | Dense 128 Dropout 0.5 | Dense 64 Dropout 0.5 | Dense 3 Softmax |

Fig. 4 CNN model for training

2.5 CNN Network Design

A CNN was created with four convolutional layers and three fully connected layers with total 922,627 trainable parameters. Each convolutional layer used 3×3 filters with '*same padding*' followed by 2×2 '*maxpooling*' layer. The output of the 4th convolutional layer was flattened followed by three fully connected layers interspaced with 50% dropout. The inner layers used *ReLU* activation, whilst the final dense layer uses *Softmax* to put the patch into one of the 3 defined classes *0 (No lesion), 1 (Red lesion)* or *2 (White lesion)* (Fig. 4).

2.6 Model Compilation and Fitting

The model was trained using 90:10 split of the training and validation data, respectively. We used '*Adam optimiser*' and '*sparse_categorical_crossentropy*' loss function to train the model. At the end of 30 epochs, the validation accuracy and loss stood at 0.889 and 0.355, respectively. Each epoch took about 15 s, and no further improvement in the validation accuracy was noticed beyond 30 epochs (Fig. 5).

3 Results

The trained CNN model was run on the 20% data reserved for testing (11290 samples). The overall accuracy was 0.906 with an average processing time of 172 μs per sample. 90% of no lesions, 86% red lesions and 94% of white lesions were

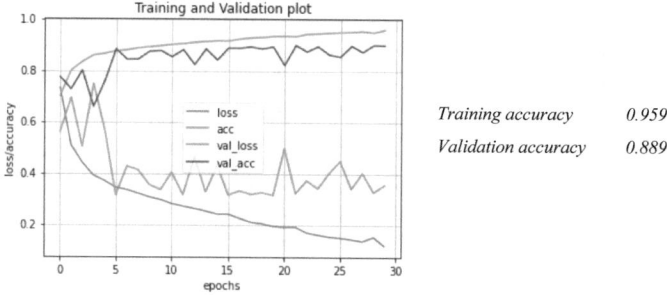

Training accuracy	0.959
Validation accuracy	0.889

Fig. 5 Training and validation plot

classified correctly in Table 4. In medical domain, more than overall accuracy, it is important to refer Table 5 to the other performance measures like *Sensitivity, Specificity, Positive Predictive Value and Negative Predictive Value*. For this, we consider the presence of either red lesion or white lesion as a positive case and no lesion as negative case.

Table 4 Category-wise classification accuracy

Category	Detected	Actual	Accuracy
No lesions	2776	3084	0.90
Red lesions	3061	3558	0.86
White lesions	4389	4648	0.94
Total	10226	11290	0.90

Table 5 Classification performance measures

Measure	Formula	Value
Sensitivity	$Sn = \frac{TP}{TP+FN}$	0.96
Specificity	$Sp = \frac{TN}{TN+FP}$	0.90
PPV	$PPV = \frac{TP}{TP+FP}$	0.96
NPV	$NPV = \frac{TN}{TN+FN}$	0.88

True Positive (TP): If the CNN classifies an actual red/white lesion as either red or white
True Negative (TN): If the CNN classifies an actual no lesion as no lesion
False Positive (FP): If the CNN classifies an actual no lesion as either red or white
False Negative (FN): If the CNN classifies an actual red/white lesion as no lesion

4 Discussion and Conclusions

In this study, we proposed a CNN model for reliably detecting red and white lesions in retinal fundus images, which are the clinically significant features used by trained ophthalmologists to diagnose and grade diabetic retinopathy. The proposed method to detect 64×64 patches around the red and white lesions yielded high sensitivity (0.96) and specificity (0.90). Sensitivity is the ability of the model to classify the positive cases correctly and specificity the ability to identify the negative cases correctly. This is a marked improvement over previous works [13–15] on same data from RIO using conventional image processing and SVN machine learning methods.

One obvious advantage of deep CNNs is that little or no 'domain knowledge' is needed. There is no need to detect or specify any individual feature, absolute or relational and no human intervention required in the training process. DCNN approaches for final DR grading based on the input image and final DR grade like [19] tend to produce very high results. Despite the high accuracy, a fully automated system will never replace the human expert in the medical domain. It will always be a 'Computer-Assisted System', than an automated system. Therefore, it becomes all the more important for the method not only to analyse and grade the input, but also visually represent to the medical expert the basis for it to reach such a conclusion, like marking out the detected lesions in the image to be reviewed, which is lacking in [19], whereas [20] and [21] present a DCNN approach for detecting micro-aneurisms. However, the other clinically relevant lesions like dot/blot/flame haemorrhages, hard exudates and cotton wool spots were not addressed in either method. Our method detects all red and white lesions and achieved better results than both these methods.

Further work is proposed to run this CNN model on a full fundus image, breaking it up into 64×64 patches and train a network to classify the whole image as various grades of DR based on the detected red and white lesions. The intention is to employ the proposed method for diabetic retinopathy screening in diabetic clinics and peripheral hospitals where the services of a qualified ophthalmologist will be scarce. Only, cases detected by the method as 'positive' need to be referred to the ophthalmologist. Studies have shown that about 17% of diabetics in India have some stage of DR, whilst the population to ophthalmologist ratio is about 1:80000. A high percentage of false positives will make the system economically unviable. On the other side, false negatives may go unattended by the expert and thereby that patient is denied the required medical attention. The PPV value of 0.96 means that 96% of the cases classified as positive by our method are actually positive, and only, 4% of it is false positives. At the same time, NPV of 0.88 means 88% of cases classified by the method as negative is actually negative and doesn't get referred to the ophthalmologist significantly reducing his workload.

Acknowledgements The authors like to acknowledge the support received from the team of ophthalmologists from the Regional Institute of Ophthalmology in providing the annotations and grading of the fundus images. Acknowledgements are also due to the Ministry of Electronics and Information Technology, Govt. of India for partially sponsoring the study and the management of the Centre for Development of Advanced Computing, Thiruvananthapuram, for facilitating the data.

References

1. Mariotti SP, Pascolini D (2010) Global estimates of visual impairment: 2010. Br J Ophthalmol 96(5):614–618
2. http://www.diabetesatlas.org
3. Deepa M, Pradeepa R, Rema M, Mohan A, Deepa R, Shanthirani S, Mohan V (2003) The Chennai urban rural epidemiology study (CURES). JAPI 51
4. Scotland GS (2007) Cost-effectiveness of implementing automated grading within the national screening programme for diabetic retinopathy in Scotland. Br J Ophthalmol 91(11):1518–1523
5. Bhaskaranand M (2015) EyeArt + EyePACS: Automated retinal image analysis for diabetic retinopathy screening in a telemedicine system. Proceedings of the ophthalmic medical image analysis second international workshop, OMIA, Munich, Germany, pp 105–112
6. Walton OB (2016) Evaluation of automated teleretinal screening program for diabetic retinopathy. JAMA Ophthalmol 134(2)
7. Roy R, Sharma T, Lob A, Pal B, Oliveira C, Raman R (2014) Automated diabetic retinopathy imaging in Indian eyes: A pilot study. Indian J Ophthalmol 62(12)
8. Fleming AD, Philip S, Goatman KA, Olson JA, Sharp PF (2006) Automated microaneurysm detection using local contrast normalization and local vessel detection. IEEE Trans Med Imaging 25(9):1223–1232
9. Goatman K (1997) Automated detection of microaneurysms. Report in Bio-Medical Physics and Bio-Engineering, University of Aberdeen. 1–8
10. Bhalerao A, Patanaik A, Anand S, Saravanan P (2008) Robust detection of microaneurysms for sight threatening retinopathy screening. In Proceeding of 6th Indian conference on computer vision, graphics and image processing. ICVGIP, pp 520–527
11. Hatanaka Y, Nakagawa T, Hayashi Y (2007) CAD scheme to detect hemorrhages and exudates in ocular fundus images. Proc SPIE 6514:1–8
12. Niemeijer M, Ginneken BV, Russell SR, Suttorp-Schulten MSA, Abràmoff MD (2007) Automated detection and differentiation of drusen, exudates, and cotton-wool spots in digital color fundus photographs for diabetic retinopathy diagnosis. Invest Ophthalmol Visual Sci 48:2260–2267
13. Sharath Kumar PN, Rajesh Kumar R, Sathar A, Sahasranamam V (2013) Automatic detection of exudates in retinal images using histogram analysis. In: Proceedings of IEEE recent advances in intelligent computational systems
14. Sharath Kumar PN, Rajesh Kumar R, Sathar A, Sahasranamam V (2014) Automatic detection of red lesions in digital color retinal images. In: Proceedings of international conference on contemporary computing and informatics, IC3I
15. Sharath Kumar PN, Deepak RU, Sathar A, Sahasranamam V, Rajesh Kumar R (2016) Automated detection system for diabetic retinopathy using two field fundus photography. 6th International conference on advances in computing and communications
16. Pratt H, Coenen F, Broadbent DM, Harding SP, Zheng Y (2016) Convolutional neural networks for diabetic retinopathy. International conference on medical imaging understanding and analysis, Loughborough
17. Gulshan V (2016) Development and validation of a deep learning algorithm for detection of diabetic retinopathy in retinal fundus photographs. J Am Med Assoc 316(22):2402–2410
18. García G, Gallardo J, Mauricio A, López J, Del Carpio C (2017) Detection of diabetic retinopathy based on a convolutional neural network using retinal fundus images. In: Lintas A, Rovetta S, Verschure P, Villa A (eds) Artificial neural networks and machine learning–ICANN 2017. Lecture notes in computer science, vol 10614. Springer, Cham. https://doi.org/10.1007/978-3-319-68612-7_72
19. Shankar K, Zhang Y, Liu Y, Wu L, Chen C (2020) Hyperparameter tuning deep learning for diabetic retinopathy fundus image classification. IEEE Access 8:118164–118173. https://doi.org/10.1109/ACCESS.2020.3005152

20. Eftekhari N, Pourreza HR, Masoudi M, Ghiasi-Shirazi K, Saeedi E (2019) Microaneurysm detection in fundus images using a two-step convolutional neural network. BioMed Eng OnLine 18:67. https://doi.org/10.1186/s12938-019-0675-9
21. Zhang L, Feng S, Duan G, Li Y, Liu G (2019) Detection of microaneurysms in fundus images based on an attention mechanism. Genes 10:817

Predicting IMDB Movie Ratings Using RoBERTa Embeddings and Neural Networks

Anagha Jose and Sandhya Harikumar

1 Introduction

In this work, the researchers obtained the data set from the IMDb website [1]. This data set contains 85,855 movies and includes information such as movie information, reviews, votes, genre data, and so on. The number of films released each year increases. Some of the films were flops at the box office, while others are box office hits. Based on a range of characteristics, the researchers developed a model to estimate the success and failure of upcoming films. The proposed approach would predict whether the film would be a success or a disaster before it is released. There are exactly the same number of positive and negative reviews in the sample. Only reviews that have the ability to polarise are given serious consideration. If a review has a score of 4 or below, it is considered negative. If a review has a score of 7 or above, it is considered positive. There is more unannotated data in the data set. IMDb users should be able to vote on a 1–10 scale, regardless of whether they consider the movie they are voting on good or bad.

RoBERTa [2] is a transformer encoder stack that has been trained. Self-attention and a feed-forward network are included in each encoder layer. The encoder's inputs pass through a self-attention layer, which then passes the results to a feed-forward network, which finally passes them on to the next encoder. In contrast to the BERT model, the RoBERTa model fine-tunes the original BERT model, which incorporates data and input manipulation. As a result, RoBERTa has had extensive training on a data set that has 160 GB of uncompressed text.

A. Jose (✉) · S. Harikumar
Department of Computer Science and Engineering, Amrita Vishwa Vidyapeetham,
Amritapuri, India
e-mail: anaghajose13@gmail.com

S. Harikumar
e-mail: sandhyaharikumar@am.amrita.edu

J. Mathew et al. (eds.), *Responsible Data Science*, Lecture Notes
in Electrical Engineering 940, https://doi.org/10.1007/978-981-19-4453-6_13

Internet movie database (IMDb) [1] is an online database of information related to films, television programmes, home videos, video games, and streaming content. In addition to this, it has extensive information on the cast, production crew, plot summaries, ratings, and critical reviews. Anyone with an IMDb account can rate a film between one and ten. Each vote is added together to form a final IMDb rating, which appears on the film's main page. The movie (or TV show) must have been screened publicly at least once before it was released. While users can make changes to their votes as frequently as they had like, a new vote overwrites any prior vote on the same title, so each user only gets one vote per title. How a user votes on a title depends on the movies and shows they have watched as well as their personal likes. Critics or other IMDb users do not have to like every movie in order for it to be considered successful. In addition, "worthless" titles are sometimes rated positively, as people who hold divergent views on such titles often aim to reduce this problem by a weighted average. All IMDb users should be able to vote on the 1–10 scale using their intuition.

1.1 Main Contributions

- Proposed a model for predicting IMDb movie rating using RoBERTa embedding and neural network.
- Only the embedding approaches of the pretrained RoBERTa model are being used here. Embeddings are linguistic modelling and feature extraction approaches, where words or phrases from the lexicon are mapped to vectors of actual numbers.
- After the RoBERTa embedding, the vector values are fed into the neural network, which predicts a value between 0 and 10.
- Also demonstrated through experiments that the neural network model outperforms the performances obtained by combining the RoBERTa embedding method with the other machine learning models.

2 Related Works

In this work, the researchers used data sets from the IMDb website [1]. The movie data set contains 85,855 movies with parameters like movie description, average rating, number of votes, genre, and so on. Some of the films were flops at the box office, while others were box office hits. Based on a range of characteristics, the researchers developed a model to estimate the success and failure of upcoming films.

Keerthi Kumar [3] has used a hybrid feature extraction method to build a sentiment analysis on IMDb movie reviews. The IMDb movie reviews are one of several data sources used in the creation of the sentiment analysis. The requirement is that the various reviews that have been gathered need to be assigned a polarity. Each of the reviews was given a polarity to be conveyed in the final output. HFEM is utilised to isolate and separate features obtained through machine learning and lexicon-based feature extraction approaches.

Melville et al. [4] investigated feature extraction via lexical approaches. The positive and negative word counts in the text served as a baseline for lexical knowledge, and the likelihood that a document belongs to a specific class was calculated. The use of pooling multinomial classifiers that incorporate both training sets and historical knowledge is the main contribution.

Nair [5] discussed the survey of various jobs performed on discharge summaries and the available technologies that were investigated. The discharge report contains detailed information about the patient, including his or her background, illnesses, examinations, treatments, and prescriptions. The discharge summary is organised in general, but not in a way that healthcare systems can process it. Several natural language processing (NLP) and machine learning algorithms were utilised on discharge summaries to extract various important bits of information.

Menon [6] developed a deterministic model-based method for producing semantically oriented topic representations from a text document. Create two matrices from the updated topic model: a document-topic matrix and a term-topic matrix. The reduced document-term matrix generated from these two matrices is 87% identical to the original document-term matrix, meaning that the original document collection and the documents reconstructed from the above two matrices are 87% comparable. Topic models infer themes purely by observing the recurrence of terms. However, the terms may not be semantically related in a way that is relevant to the topic.

Akhil Dev [7] suggests a model for document data structure maintenance utilising a variety of deep learning approaches. The majority of strategies may be compared based on their vector similarity. The research work contributes to enhancing results by comparing several types of word sequencing approaches used to maintain the document's structure. The collected results show that the Doc2Vec model outperforms the Tfidf model in terms of word order similarity.

Asritha [8] highlighted the benefits of intelligent text mining. Cyberbullying is described as using the Internet to abuse, harass, threaten, or humiliate another individual. In the recent years, cyberbullying has become increasingly popular on social media sites such as Facebook, YouTube, and Instagram. This can be mitigated somewhat by segregating such scary messages or comments. Sentiment analysis is the technique of determining if a sentence is good, negative, or neutral. It aids in establishing the emotional tone of a sentence. This article describes a hybrid classifier technique for categorising these frightening messages, which identifies reviews as good or negative. According to the experimental results, the classifier is 89.36% accurate for the data set under consideration.

The feature extraction and processing system [9] are designed to extract information from English news items and deliver it to the user in a logical manner. Crawl and save information about a predefined collection of websites. News organisations conceal a great deal of information within the pages of their publications. Extracting data and organising it in a way that allows for conclusions are critical in analytics.

The application pulls identifying components from news articles, such as the place, person, or organisation mentioned, as well as the title and important phrases.

The semantic representation of documents based on matrix decomposition [10] focuses on the crucial problem of semantic text classification for data gathering in a data implementation. Frequently, search requests on documents seek out essential data. Standard feature extraction algorithms do not convey significance and instead focus on phrase similarity when processing queries. The difficulty that semantic information in texts introduces is in distinguishing essential aspects. The bulk of techniques for discovering relevant traits alters the original data by transforming it into another space. This culminates in a sparse matrix that is expensive to compute. This technique optimises data aggregation by identifying key documents and phrases. Experiments using five sets of data indicate the strategy's efficacy.

3 Proposed Method

The architecture for predicting IMDb movie ratings using RoBERTa embeddings and a neural network is depicted in Fig. 1 [11]. The model is made up of a RoBERTa server, which encodes the text and returns an embedding vector, and a neural network model, which predicts a value between 0 and 10 (rating). The collected IMDb movie data set is preprocessed, free of unwanted data, space, and missing values, and then fed into the RoBERTa server. Each component's specific design and implementation is discussed further below.

3.1 RoBERTa

RoBERTa [2] is an improved BERT pretraining approach. RoBERTa exceeds BERT by making the following changes:

- BookCorpus (16 G), CC-NEWS (76 G), OpenWebText (38 G), and Stories (31 G) data are used by RoBERTa, but BERT exclusively uses BookCorpus for training.
- For the Masked Language Model (MLM) aim, BERT masks training samples once, whereas RoBERTa replicates training sets 10 times and masks it differently.

This model was created and published by Facebook. This pretrained word representation language model can process long texts efficiently. It is bidirectional, so it can be read from both directions. Special tokens are used: $\langle s \rangle$, $\langle /s \rangle$, [MASK].

The RoBERTa model did not contain the next sentence prediction (NSP) task that we use in the BERT pretrained model. For example, if we have phrases A and B, the model will try to determine whether phrase B follows phrase A. The RoBERTa model represents dynamic masking, so that during training times, the masked token changes. In the training procedure, larger lots of training sizes have also been found

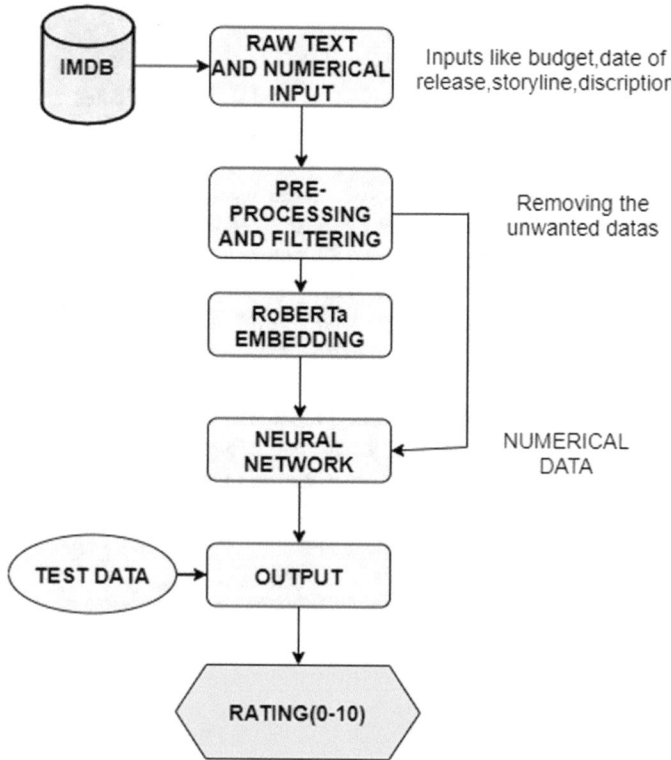

Fig. 1 Block diagram of proposed method

to be more helpful. The authors tested the removal and addition of NSP loss and found that removing NSP loss improves the model task performance, possibly a bit. This RoBERTa model can process long text, but it is very expensive, so we can only use a pretrained model.

3.2 Data set

In this work, the researchers obtained the data set from the IMDb website [1]. The collection contains 85,855 movies, each with its own definition, overall score, number of votes, genres, and so on. Each year, the number of films released increases. Some of the films are box office flops, while others are box office hits. Internet movie database (IMDb) [1] is a website that contains information about movies, television programmes, video clips, games consoles, video streaming, and other streaming content.

3.3   RoBERTa Embeddings

The RoBERTa model [2] uses the MLM, a model that replicates and masks the training set ten times and predicts a masked word's original value. Transform paper paragraphs into embedding with the RoBERTa pretrained model (a robustly optimized BERT pretraining approach). The model uses token embeddings, segment embeddings, and position embeddings for the conversion of embeddings. The model adds the tokens in token embeddings, and the number of the sentence that is encoded into a vector in segment embeddings, and the position of a word within that sentence that is encoded into a vector in position embeddings. These values are concatenated by the model. The RoBERTa is used to extract features from text data, such as word and sentence embedding vectors. The RoBERTa model is retrieved from the Internet by using the RoBERTa-base-uncased package. The RoBERTa transformer is a stack encoder. The model contains a neural feed network and self-attention in each encoder. The concept of attention has helped neural machine translation programmes perform better.

3.4   Neural Network

Artificial neural networks [11] (also known as deep neural networks or DNNs) belong to a family of machine learning techniques based on learning data representations. In this context, the definition of a DNN is represented as a function $F: X \rightarrow Y$, with X denoting the input field and Y denoting the output field. It is made up of multiple layers connected to one another by connecting links that have different weights associated with them. Once a DNN has been trained, its training phase involves identifying the numerical values of the weight matrices. To quantify the quality of the training operation, a loss function is employed to measure the discrepancies between the training targets and the actual labels. During the training phase, the parameters are updated using the backpropagation approach to minimise the loss function.

4   Experiments and Results

The algorithm forecasts whether the film will be a hit or a flop in the current work. The researchers estimated movie ratings using robustly optimized BERT pretraining approach (RoBERTa) [2] embeddings and a neural network model. Prior to employing RoBERTa, text input must be translated to numeric token identifiers and stored in several tensors. Using the neural network model, we will attempt to compare the accuracy performance of the model with the accuracy performance of the combinations of the RoBERTa embedding method with the machine learning models, which are the logistic regression model, the random forest model, and the support vector

Table 1 Comparison of RoBERTa embedding with neural network and other machine learning models

No. of labelled texts	70	80	90	100
Neural network	85.79	85.49	85.40	85.95
Logistic regression	71.05	71.52	71.69	72.47
Random forest	56.77	63.77	64.42	63.49
Support vector machine	68.58	68.23	68.82	71.61

```
loss, accuracy = classifier_model.evaluate(test_ds)

print(f'Loss: {loss}')
print(f'Accuracy: {accuracy}')

782/782 [==============================] - 1834s 2s/step - loss: 0.4697 - binary_accuracy: 0.8514
Loss: 0.469670832157135
Accuracy: 0.851360023021698
```

Fig. 2 The accuracy and loss of the model

machine (SVM) model. Google Colab is equipped with an NVIDIA GPU and 12 GB RAM, which allows us to run our models. Graph neural network model is at least as good as combinations of RoBERTa embedding method with logistic regression model, random forest model, and support vector machine model, as shown in the preceding Table 1, but it often outperforms them in terms of accuracy performance because the neural network model works perfectly even when the labelled text is sparse (as shown in the preceding Table 1).

The first layer is the embedding layer. This layer searches for an embedding vector for each word-index in the integer-encoded reviews. These vectors are used to train the model. The vectors increase the dimension of the output array. A Global-AveragePooling1D layer produces a fixed-length output vector for each sample by averaging across the sequence dimension. This simplifies the model's handling of inputs of different lengths. This fixed-length output vector is routed through a neural network layer with 16 hidden units. The last layer is composed of a single densely connected output node. Because this is a binary classification task and the model outputs a probability, and the model applies the losses-binary crossentropy loss function (a single-unit layer with sigmoid activation). Experiment proves that the neural network model outperforms the performances obtained by combining the RoBERTa embedding method with the other machine learning models. This straightforward approach achieves an accuracy of approximately 85% (Fig. 2).

5 Conclusion

In the current work, the model is used to forecast whether a film will be a hit or a disaster. Several of the films have flopped at the box office, while others have been box office hits. The researchers developed a model to anticipate the success and failure of forthcoming films using a range of variables. The determinants include the date and month of release, the cast and director, the narrative and genre, and the language. Prior to the film's release, the suggested model will forecast its success or failure. To estimate movie ratings, the researchers employed RoBERTa embeddings and a neural network model that predicted a value between 0 and 10 (rating). Due to the fact that manufacturers incur significant losses as a result of investment, this technique benefits them. This will assist the business in earning as much profit as feasible. The neural network model outperforms other machine learning models when combined with the RoBERTa embedding approach, according to the results of the experiments.

Internet movie database (IMDb) [1] is an online database of information about movies, TV programmes, video clips, games consoles, and streaming entertainment. In addition to this, it has extensive information on cast, production crew, plot summaries, ratings, and critical reviews. If a review has a score of 4 or below, it is considered negative. If the predicted rate of the movie has a score of 7 or above, it is considered positive. Future studies could include assessing a semantic search engine employing OpenAI's generative pretrained transformer (GPT) models. These GPT models are some of the most powerful language models available. These models are capable of performing a variety of NLP tasks, such as question answering, textual entailment, and text summarisation among others.

References

1. IMDB Webpage (2007) Dataset. https://www.imdb.com/interfaces/
2. Devlin J, Chang M-W, Lee K, Toutanova K (2018) BERT: pre-training of deep bidirectional transformers for language understanding. arXiv preprint arXiv:1810.04805
3. Keerthi Kumar HM, Harish BS, Darshan HK (2018) Sentiment analysis on IMDb movie reviews using hybrid feature extraction method. Int J Interact Multimedia Artif Intell (2018). https://doi.org/10.9781/ijimai.2018.12.005
4. Melville P, Gryc W, Lawrence RD (2009) Sentiment analysis of blogs by combining lexical knowledge with text classification. In: Proceedings of the 15th ACM SIGKDD international conference on knowledge discovery and data mining. ACM, pp 1275–1284
5. Nair PC, Gupta D, Devi BI (2021) A survey of text mining approaches, techniques, and tools on discharge summaries. In: Gao XZ, Tiwari S, Trivedi M, Mishra K (eds) Advances in computational intelligence and communication technology. Advances in intelligent systems and computing, vol 1086. Springer, Singapore. https://doi.org/10.1007/978-981-15-1275-9-27
6. Menon RRK, Joseph D, Kaimal MR (2017) Semantics-based topic inter-relationship extraction. J Intell Fuzzy Syst 32(4):2941–2951
7. Menon RRK, Akhil Dev R, Bhattathiri SG (2020) An insight into the relevance of word ordering for text data analysis. In: 2020 fourth international conference on computing methodologies and communication (ICCMC), pp 207–213. https://doi.org/10.1109/ICCMC48092.2020.ICCMC-00040

8. Asritha P, Reddy PPR, Sudha CP, Neelima N (2021) Intelligent text mining to sentiment analysis of online reviews. In: ICASISET.EA. https://doi.org/10.4108/eai.16-5-2020.2303907

9. Karumudi GVNSK, Sathyajit R, Harikumar S (2019) Information retrieval and processing system for news articles in English. In: 2019 9th international conference on advances in computing and communication (ICACC), pp 79–85. https://doi.org/10.1109/ICACC48162.2019.8986223

10. Baladevi C, Harikumar S (2018) Semantic representation of documents based on matrix decomposition. In: International conference on data science and engineering (ICDSE), pp 1–6. https://doi.org/10.1109/ICDSE.2018.8527824

11. Patel M (2019) TinySearch—semantics based search engine using BERT embeddings. Twitter as a tool for the management and analysis of emergency situations: a systematic literature review. Int J Inf Manag 43:196–208. arXiv preprint arXiv:1908.02451

Domain-Specific Type-Safe APIs for Hierarchical Scientific Data with Modern C++

William F. Godoy ⓘ, Addi Malviya Thakur ⓘ, and Steven E. Hahn ⓘ

1 Introduction

Scientific data products from simulations, observations and experiments are represented either in-memory or as files using a multitude of text or binary data formats. Hierarchical, self-describing (combining data and metadata) and array-oriented data formats are broadly applicable as they provide extensible schema-based interfaces to store and interact with scientific data [23]. Popular file formats are: the Hierarchical Data Format, HDF5 [26], the Network Common Data Form, NetCDF [19] (which depends on HDF5), while others tackle specific domains, such as FITS [30] for astronomy applications, and those with added streaming capabilities such as: ROOT in high-energy physics [1], and ADIOS [8] in high-performance computing (HPC) applications. There is a plethora of self-describing hierarchical schemas defined according the needs of each scientific community, which rely upon the underlying file format for data archival and sustainability [4]. In addition, as interoperability becomes a requirement, for example the HDF5 Virtual Object Layer (VOL) [2] and robust APIs are required to address the differences between data model paradigms [17].

The United States Government retains and the publisher, by accepting the article for publication, acknowledges that the United States Government retains a nonexclusive, paid-up, irrevocable, world-wide license to publish or reproduce the published form of this manuscript, or allow others to do so, for United States Government purposes. The Department of Energy will provide public access to these results of federally sponsored research in accordance with the DOE Public Access Plan (http://energy.gov/downloads/doe-public-access-plan). Work at Oak Ridge National Laboratory was sponsored by the Division of Scientific User Facilities, Office of Basic Energy Sciences, US Department of Energy, under Contract no. DE-AC05-00OR22725 with UT-Battelle, LLC.

W. F. Godoy (✉) · A. M. Thakur · S. E. Hahn
Computer Science and Mathematics Division, Oak Ridge National Laboratory, Oak Ridge, TN 37830, USA
e-mail: godoywf@ornl.gov
URL: https://csmd.ornl.gov

This is a U.S. government work and not under copyright protection in the U.S.; foreign copyright protection may apply 2022
J. Mathew et al. (eds.), *Responsible Data Science*, Lecture Notes in Electrical Engineering 940, https://doi.org/10.1007/978-981-19-4453-6_14

191

Findable, Accessible, Interoperable and Reproducible (FAIR) [31] data has become a requirement for modern software development for reproducible computational sciences. Hinsen and Läufer [14] argue that source code is the only precise and rigorous specification for a computational scientist and have become the equivalent of mathematical formulas in theoretical science. In addition, they propose that while traditional computer science has focused on algorithms and software, data is the most relevant aspect to a scientific domain. Data must remain usable for many years to come, thus requiring documentation of all formats independently of the source code. According to Gray et al. [12], the last decades have been marked by the new paradigm of data-intensive science. Scientific data management has become challenging as the complexity of the "5Vs" of big data: Volume, Velocity, Variety, Veracity and Value increases [15]. As such, data search and extraction have become even more relevant in order to access quantities of interests in an efficient manner [11]. Therefore, there is a strong need to find new interaction models that ease the access of available large and complex scientific datasets.

In this paper, we propose to incorporate three metaprogramming features available in the modern C++17 standard: (i) template `auto` type deduction [27], (ii) valid template arguments (e.g. `enum class`) for user-defined types and (iii) traditional preprocessor macro for templated code generation. We use these features to design type-safe APIs for well-defined self-describing hierarchical data. These APIs facilitate interactions for multidimensional array data access and "in-memory" indexing generation. Our approach is expected to provide guarantees that certain kinds of exceptions never occur in these data structures, thereby increasing compile-type safety. As mentioned by Tratt [28], the latter is a desirable feature in programming models for domain-specific applications. We demonstrate this on experimental raw neutron scattering data mapped and implemented using the HDF5 data model. The structure of the remainder of the paper is as follows: Sect. 2 provides a description of related work, Sect. 3 provides a brief overview of the hierarchical self-describing model used in scientific applications, specifically the HDF5 data model, and the generation of "in-memory" indices. Section 4 introduces the proposed type-safe C++17 APIs, while Sect. 5 shows their application on a real use-case based on experimental neutron science data including a discussion on potential trade-offs. Finally, Sect. 6 presents the summary of our findings and APIs pros and cons, while Sect. 7 provides the conclusions of this work.

2   Background

Generic programming has become a widespread paradigm in the creation of highly adaptable and critical software libraries [25]. C++ templates [24] provide a Turing-complete mechanism, which includes type-safety at compile-time, that has evolved with recent standards of the language. There have been few attempts to exploit C++ templates to provide type-safety in scientific datasets. For the most part, data interaction is the task of the input/output I/O layer of an application-specific software stack to facilitate communication across scientific data stakeholders using a well-defined schema, as described in the large number of users of the HDF5 data model [4]. A recent effort to add type-safety on top of HDF5 using C++ is the h5cpp header-only library [29]. h5cpp connects the underlying runtime HDF5 library with well-defined data models of popular C++ templates-based linear algebra libraries, such as Armadillo [21] and Eigen [13]. Golasowski et al. [10] proposed an automated code generation phase, to avoid the error-prone handwritten task of data manipulation, for each compound type using a DSL approach. The result is the ANTAREX DSL [22] for HPC applications. Gaur et al. [5] provided a C++ template-based approach to store satellite image data using HDF5. The present work aims to leverage efforts related to type-safe data interactions and current features of the C++ template generic programming only available in the latest C++17 standard without the need to resort to a full DSL.

3 Self-describing Hierarchical Scientific Data

This section describes briefly the hierarchical model used by the HDF5 library to store scientific data. While we focus on HDF5, the scope of this work could benefit other scientific data frameworks that follow a similar data model.

3.1 Hierarchical Data Model

Figure 1 shows a schematic basic representation of the HDF5 data model from [26]. At the highest level a HDF5 file is a container of "Groups", which determine the hierarchy used to represent the different "Dataset" array-based objects stored within a scientific application. "Attribute" objects contain additional metadata annotations that relate to a particular "Group" or "Dataset" object.

As a result, the typical hierarchical structure resembles the structure of a Unix file system [20]. A flat hierarchy representation is given in Table 1, in which the root level and a hierarchy sequence are denoted by the slash character (/). Interactions are determined by two important aspects: programming model APIs and in-memory indexing.

Programming Model APIs: Traditional HDF5 and binding libraries have been of a runtime characteristics for interacting with each component due to their general-purpose nature. As shown in Listing 1.1, datasets must be identified by character arrays in write operations.

Listing 14.1 General-Purpose HDF5 C API for writing datasets

```
hid_t H5Dopen( hid_t loc_id, const char *name )
herr_t H5Dwrite( hid_t dataset_id, ... )
```

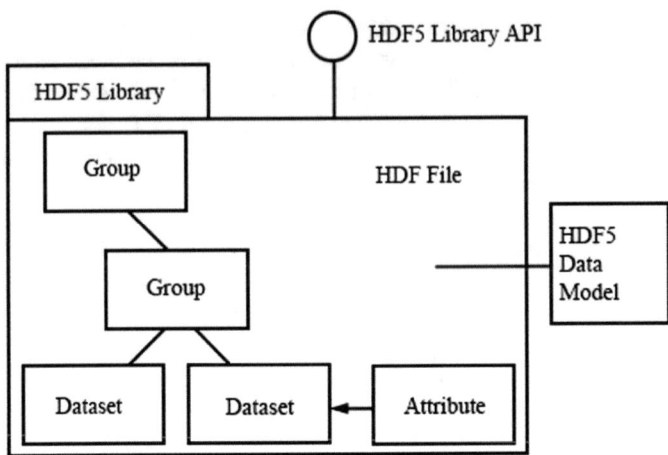

Fig. 1 Schematic representation of the HDF5 data model used in APIs illustrating the use of "group", "dataset" and "attribute" objects for interacting with self-describing array-based scientific data from [26]

Table 1 Schematic flat representation of the hierarchy in a HDF5 data model showing group, dataset and attribute objects as the basic components

Data type	Entry name
Group	/group1
Attribute	/group1/attribute
Dataset	/group1/dataset
Attribute	/group1/dataset/attribute
Group	/group1/group2
Attribute	/group1/group2/attribute
Dataset	/group1/group2/dataset
Attribute	/group1/group2/dataset/attribute

In-memory Indexing: It refers to the representation of the metadata index that is constructed to align to the application needs for data search and processing. In self-describing data formats, indexing is highly efficient as the data and metadata are part of the same stream [32]. There is no standard way to construct an index in-memory from the stored metadata, and different data structures (e.g. binary trees, tries, and hash tables) must be carefully studied. This is an active area of research [32], which includes techniques such as queries [11]; mainly due to the impractical nature of a "one-size-fits-all" standard solution that balances data representation with input output (I/O) bandwidth access [4, 18, 33].

4 Metaprogramming Type-Safe C++ APIs

4.1 Template Type Auto Deduction with C++17

The first step is to understand the role of the template auto deduction feature [27] introduced in the C++17 standard. Listing 1.2 shows how a template argument "type" can be auto-resolved at compile-time. Thus resulting in simpler APIs, since it is possible to pass a non-type value argument to a template parameter.

Listing 14.2 C++17 template <*auto*> feature example.

```
// prior to C++17 types must be explicit:
template<class Type, Type value>
void Foo()

// with C++17 auto public facing API is simplified
// explicit type implementation is internal
template <auto value>
void Foo(){ FooImpl<decltype(value), value>(); }

template <class Type, Type value>
void FooImpl(){ ... }

// Usage: prior to C++17
Foo<int, 5>();
Foo<float, 10>();

// Usage: with C++17 type is auto deduced
Foo<5>();
Foo<10.f>();
```

Combining template type auto deduction along with the strongly-typed enum class feature introduced in C++11 provides the possibility of user-defined types to be template arguments. Listing 1.3 illustrates the generic application of these concepts to the hierarchical data model described in Sect. 3. The result is a simplified and type-safe API that maps C++ nested namespaces to hierarchical data semantics in scientific formats (/).

Therefore, a preliminary mapping of the HDF5 data model to C++ valid template arguments can be established as summarized in Table 2. While this mapping is only referential, it provides an initial path for connecting well-defined deep hierarchy schemas with more programmatic and type-safe APIs. Some drawbacks of mapping flexible "string" representations with C++ types include conflicts with restricted keywords and special characters that are part of the language syntax. These limitations are illustrated in our example use-case in Sect. 5.

Listing 14.3 Type-safe C++17 API for hierarchical entry schematic example

```cpp
// Generic interface for data interaction
template<class T>
void Foo(const std::string& entry, const T* data);

template<class T>
T FooIndex();

// Usage:
Foo("/group/variable", data);
const std::set<std::string> index = FooIndex();

// Type-safe API replaces runtime string with type
template <auto entry, class T>
void Bar(const T* data){
    BarImpl<decltype(entry), entry>(data); }

template <auto model, class T>
T BarIndex();

// Entry definition as an enum class
enum class group{ variable }

// Specialization definition
template Bar<group::variable, int>(const int* );

// Usage of C++17 type-safe API
Bar<group::variable>(data);
const std::set<std::string> flatIndex =
    BarIndex<model1, std::set<std::string>>();
const std::map<std::string, std::string> nxClassIndex =
    BarIndex<model2, std::map<std::string, std::string>>();
```

Table 2 Mapping between HDF5 data model components and valid template argument C++17 features for type-safe APIs

HDF5 data model component	C++ template argument
Group	namespace
	enum class
Attribute	int
	enum class entry
Dataset	int
	enum class
	enum class entry
Dataset attribute	int

Table 3 Schematic representation of the hierarchical NeXus schema [16] for recorded raw event-based neutron data from [9]

Data type	Entry name
Group	/entry
Attribute	/entry/NX_class
	...
Group	/entry/DASlogs
Attribute	/entry/DASlogs/NX_class $\rightarrow$ "NXlog"
Group	/entry/DASlogs/BL6:CS:DataType
Attribute	/entry/DASlogs/BL6:CS:DataType/NX_class
Dataset (SDS)	/entry/DASlogs/BL6:CS:DataType/average_value
Dataset (SDS)	/entry/DASlogs/BL6:CS:DataType/average_value_error
	...
Group	/entry/bank1_events
Attribute	/entry/bank1_events/NX_class $\rightarrow$ "NXevent_data"
Dataset (SDS)	/entry/bank1_events/event_id
Dataset (SDS)	/entry/bank1_events/event_index
	...
Group	/entry/bank91_events
Attribute	/entry/bank91_events/NX_class $\rightarrow$ "NXevent_data"
Dataset (SDS)	/entry/bank91_events/event_id
Dataset (SDS)	/entry/bank91_events/event_index

5 Use-Case: Neutron Scattering Data

We apply the concepts outlined in Sect. 4 to our domain of interest to provide type-safe APIs for accessing neutron science data. The present use-case tackles the international standard NeXus schema [16] used for storing experimental raw neutron scattering data on top of the underlying HDF5 file format [26]. We narrow the scope to our present efforts on providing type-safe APIs to NeXus specific entries as it's a good baseline for other well-defined schemas.

Table 3 illustrates a subset of the contents typical of a NeXus conformant file. Groups are annotated with the "NX_class" attribute while generic datasets are marked as scientific datasets (SDS) by default. Therefore, any "in-memory" index representation must account for classifying the entries according to their "NX_class" attribute for certain postprocessing tasks, which does not necessarily map the file hierarchy structure stored on disk [6, 9]. Thus potential APIs must be flexible enough to adapt to the processing requirements for data search.

The proposed type-safe APIs are introduced as part of our exploratory work on the open-source C++17 No Cost Input Output (NCIO) library[1] [7]. The goal of NCIO

---

[1] https://github.com/ORNL/ncio.

Listing 14.4 Resulting type-safe C++17 API for NeXus schema using NCIO

```
// Generic function to specialize:
template <auto entry, class T>
void ncio::DataDescriptor::Put(const T* data);

// Use example:
#include <ncio.h>

// to narrow the scope to a particular schema
using namespace ncio::schema::nexus;

ncio::DataDescriptor f =
    ncio.Open("file.nxs.h5", ncio::mode::Write);
f.PutAttribute<entry::NX_class>(); // "NX_entry" default
f.Put<entry::DASlogs::BL6_CS_DataType::average_value>(data);
f.Put<entry::bank1_events::event_id>(event_id);
f.Put<entry::bank91_events::event_index>(event_index);
f.Close();
```

is to serve as a front-facing implementation that provides API semantics that are as close as possible to a standard schema in a specific scientific domain. Listing 1.4 shows the intended semantics, in which specializations of the generic template `Put` and `PutAttribute` functions for dataset and attribute interactions resemble the NeXus standard schema entries.

It is important to mention that the expected type-safety applies to the data entry passed which is limited by the specific specialization of these functions caught at compile-time. Another advantage of this programmatic approach is the fact that integrated development environments (IDEs) can recognize and narrow the scope of valid types. As illustrated in Fig. 2, the Eclipse CDT IDE [3] automatically recognizes the scope and entries for the next hierarchy level in the proposed APIs. Exploiting IDEs' powerful features are desired as the development cost of tasks could scale up with the number of schema annotations and complexity of the resulting workflow.

To achieve the user-facing APIs in Listing 1.4, the NeXus schema standard entries are "registered" in the NCIO source code. As shown in Listings 1.5 and 1.6, the relevant pieces that map the entries in Table 3 are applied as arguments to the specializations of the generic `Put` and `PutAttribute` functions. The use of C++ namespaces allows for an easy modular extension to multiple schemas at compile-time. Entries for a particular schema are encapsulated within a unique namespace scope, thus avoiding conflicts in template arguments. Listing 1.5 provides an example of the limitations of mapping language types to the more flexible "string" character arrays. Special characters (e.g. :) and reserved keywords (e.g. enum) must be handled conveniently to preserve application-specific semantics in the resulting APIs.

Listing 1.6 also shows how a generative approach can be followed using traditional C preprocessor macros for schema annotations defined in a sequence. These can be extended to other patterns as definitions must be explicitly provided in a macro list.

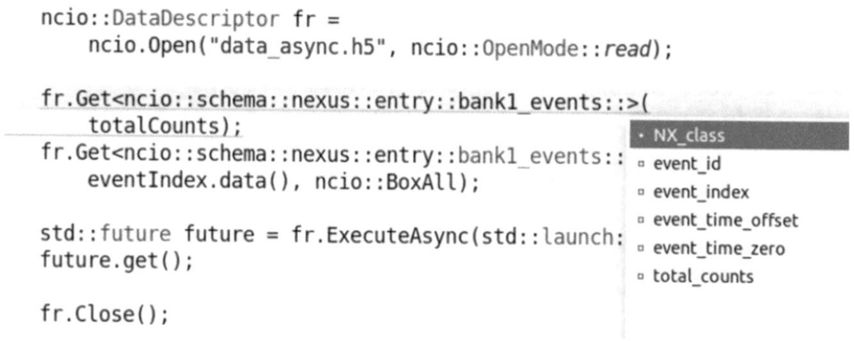

```
ncio::DataDescriptor fr =
    ncio.Open("data_async.h5", ncio::OpenMode::read);

fr.Get<ncio::schema::nexus::entry::bank1_events::>(
    totalCounts);
fr.Get<ncio::schema::nexus::entry::bank1_events::
    eventIndex.data(), ncio::BoxAll);

std::future future = fr.ExecuteAsync(std::launch:
future.get();

fr.Close();
```

```
· NX_class
□ event_id
□ event_index
□ event_time_offset
□ event_time_zero
□ total_counts
```

Fig. 2 Example for automatic recognition of type scope using the Eclipse CDT IDE [3] for the proposed type-safe API, errors are caught at development time

Listing 14.5 Template implementation in type-safe C++17 API for NeXus schema

```
// NeXus entries in extendable schema
namespace ncio::schema::nexus::entry{
// namespace -> DASLogs "group"
namespace DASLogs{
// "group" attributes
constexpr int NX_class = 0;

// "group" datasets or attributes
// replace ":" with "_" as not allowed in C++
// use "_" to differentiate from deeper hierarchies
enum class _BL6_CS_DataType_ { NX_class, average_value,
                               average_value_error };

// namespace -> BL6:CS:DataType "group"
namespace BL6_CS_DataType
{
// enum is a reserved C++ word use "_enum_"
enum class _enum_ { NX_class, name_0, name_1 };
enum class time{ offset_nanoseconds, offset_seconds};
}

} //end DASLogs
} //end ncio::schema::nexus::entry
```

The end result is a more compact representation of registered types that follow a particular pattern.

Similarly, we tackle the case for "in-memory" indexing APIs. Flexibility becomes a requirement as well-defined data types must be studied *a-priori* to match the application search and processing needs. It can be seen in Listing 1.7 that two different type-safe specializations can be implemented to obtain different indices referred as model0 and model1. In previous efforts [6, 7, 9], we concluded that a binary-tree based on the "NX_class" attribute given in Table 3, model1, is preferred for certain

Listing 14.6 Implementing type-safe APIs using MACRO for code generation

```
// NeXus entries defined in a sequence
#define NCIO_NEXUS_FOREACH_BANK_ID(MACRO)  \
    MACRO(1)                               \
    ...                                    \
    MACRO(91)

namespace ncio::schema::nexus::entry{

#define declare_nexus_bank_entry(T) \
    enum class bank##T##_events{event_id, event_index, \
                               NX_class};
NCIO_NEXUS_FOREACH_BANK_ID(declare_nexus_bank_entry)
#undef declare_nexus_bank_entry

// which expands into:
enum class bank1_events{ event_id, event_index, NX_class };
...
enum class bank91_events{ event_id, event_index, NX_class };
}
```

processing tasks, as oppose to a flat structure, model0, for performance. As more index "models" are required, the type-safety formulation in Listing 1.7 can be easily extended as needed.

6   Summary, Pros and Cons

The proposed methodology allows for type-safe extendable C++17 APIs for well-defined schemas in hierarchical scientific data, e.g. HDF5. We list the different pros and cons found in our experience.

Pros:

- Strong type-safe APIs that resemble hierarchical scientific data
- API semantics are as close as possible to the specific domain
- Schema inspection can be done with any IDE that supports auto-completion prior to compilation
- Compile-time safety and faster development benefits scale-up with the number of entries in a particular schema
- Since template types are specialized, compilation times shouldn't be affected (unlike in inlined header-only implementations).

Cons:

- Runtime requirements (e.g. index-based loops) still make the case for purely run-time APIs

Listing 14.7 Type-safe APIs and implementation for multiple "in-memory" indices

```
using namespace ncio::schema
// User-facing APIs
template<auto index, class T>
T DataDescriptor::GetMetadata();
// usage, fr is a ncio::DataDescriptor reader
const auto flatIndex =
   fr.GetMetadata<nexus::index::model0, nexus::model0_t>();
/*
/entry
/entry/NX_class
/entry/DASLogs/...
/entry/bank1_events/...
/entry/bank91_events/...
*/

const auto nxClassIndex =
   fr.GetMetadata<nexus::index::model1, nexus::model1_t>();
/*
Attribute NX_class : entries

NXentry      : /entry
NXevent_data: /entry/bank1_events
SDS          : /entry/bank1_events/event_id,
               /entry/bank1_events/event_index
               ...
               /entry/bank91_events/event_id,
               /entry/bank91_events/event_index
*/

// Internal specialized template argument
// definitions in NCIO
namespace ncio::schema::nexus
{
enum class index { model0, model1 };
// flat index
using model0_t = std::set<std::string>;
// NX_class based binary-tree
using model1_t =
    std::map<std::string, std::set<std::string>>;
}
```

- Entries might not map 1-to-1 due to C++ reserved keywords
- Added cost to the developer in registering domain-specific schema entries either manually or via automation
- APIs using move semantics and return types still require explicit template resolution (auto can't be deduced at compile-time).

7 Conclusions

We proposed the use of features available in modern C++ to develop type-safe, compile-time checked application programming interface (API) for self-describing, hierarchical scientific data which covers a broad range of applications. By combining modern C++ template metaprogramming and traditional macro for generative code, consumers of hierarchical data libraries (e.g. HDF5 and NetCDF) can access type-safe domain-specific APIs. In addition, the proposed APIs allow for more compile-time, and integrated development environment (IDE) checks to improve findability of their data. We showcase the application of these ideas and benefits to access the data and form appropriate search "in-memory" index annotations for processing. We tested the proposed concept using functional software targeting experimental neutron science data using the standard metadata-rich NeXus format on top of the HDF5 file format. While the case for "compile-time" safe APIs applies to traditional languages like C, C++ or Fortran used in scientific computing; we expect that these "type-safety" concepts can be extended to "just-in-time" dynamic languages, such as Python, Julia and R, so potential errors can still be caught early in the development process. Overall, modern capabilities of computational science must continue to be researched and provided as requirements on the quality, reliability and reproducibility of research software and FAIR scientific data increase.

Acknowledgements Work at Oak Ridge National Laboratory was sponsored by the Division of Scientific User Facilities, Office of Basic Energy Sciences, US Department of Energy, under Contract no. DE-AC05-00OR22725 with UT-Battelle, LLC.

References

1. Antcheva I, Ballintijn M, Bellenot B, Biskup M, Brun R, Buncic N, Canal P, Casadei D, Couet O, Fine V, Franco L, Ganis G, Gheata A, Maline DG, Goto M, Iwaszkiewicz J, Kreshuk A, Segura DM, Maunder R, Moneta L, Naumann A, Offermann E, Onuchin V, Panacek S, Rademakers F, Russo P, Tadel M (2009) ROOT—a C++ framework for petabyte data storage, statistical analysis and visualization. Comput Phys Commun 180(12):2499–2512. https://doi.org/10.1016/j.cpc.2009.08.005. 40 YEARS OF CPC: A celebratory issue focused on quality software for high performance, grid and novel computing architectures
2. Byna S, Breitenfeld MS, Dong B, Koziol Q, Pourmal E, Robinson D, Soumagne J, Tang H, Vishwanath V, Warren R (2020) ExaHDF5: delivering efficient parallel I/O on exascale computing systems. J Comput Sci Technol 35(1). https://doi.org/10.1007/s11390-020-9822-9
3. Eclipse Foundation: Eclipse CDT (C/C++ development tooling). https://www.eclipse.org/cdt/
4. Folk M, Heber G, Koziol Q, Pourmal E, Robinson D (2011) An overview of the HDF5 technology suite and its applications. In: Proceedings of the EDBT/ICDT 2011 workshop on array databases, AD '11. Association for Computing Machinery, New York, NY, USA, pp 36–47. https://doi.org/10.1145/1966895.1966900
5. Gaur N, Mahammad S, Sharma V, Dhar D, Ramakrishnan R (2012) Template based HDF5 satellite digital data product generation software. In: 2012 international conference on data science and engineering (ICDSE), pp 160–165. https://doi.org/10.1109/ICDSE.2012.6281903

6. Godoy WF, Peterson PF, Hahn SE, Billings JJ (2020) Efficient data management in neutron scattering data reduction workflows at ORNL. In: 2020 IEEE international conference on big data (big data), pp 2674–2680. https://doi.org/10.1109/BigData50022.2020.9377836
7. Godoy WF, Peterson PF, Hahn SE, Hetrick J, Doucet M, Billings JJ (2020) Performance improvements on SNS and HFIR instrument data reduction workflows using Mantid. In: Nichols J, Verastegui B, Maccabe AB, Hernandez O, Parete-Koon S, Ahearn T (eds) Driving scientific and engineering discoveries through the convergence of HPC, big data and AI. Springer, Cham, pp 175–186
8. Godoy WF, Podhorszki N, Wang R, Atkins C, Eisenhauer G, Gu J, Davis P, Choi J, Germaschewski K, Huck K, Huebl A, Kim M, Kress J, Kurc T, Liu Q, Logan J, Mehta K, Ostrouchov G, Parashar M, Poeschel F, Pugmire D, Suchyta E, Takahashi K, Thompson N, Tsutsumi S, Wan L, Wolf M, Wu K, Klasky S (2020) ADIOS 2: the adaptable input output system. A framework for high-performance data management. SoftwareX 12:100561. https://doi.org/10.1016/j.softx.2020.100561
9. Godoy WF, Savici AT, Hahn SE, Peterson PF (2021) Efficient loading of reduced data ensembles produced at ORNL SNS/HFIR neutron time-of-flight facilities. In: (2021) IEEE International Conference on Big Data (Big Data), 2021, pp. 2949–2955. https://doi.org/10.1109/BigData52589.2021.9671354
10. Golasowski M, Bispo J, Martinovič J, Slaninová K, Cardoso JMP (2017) Expressing and applying C++ code transformations for the HDF5 API through a DSL. In: Saeed K, Homenda W, Chaki R (eds) Computer information systems and industrial management. Springer, Cham, pp 303–314
11. Gosink L, Shalf J, Stockinger K, Wu K, Bethel W (2006) HDF5-FastQuery: accelerating complex queries on HDF datasets using fast bitmap indices. In: 18th international conference on scientific and statistical database management (SSDBM'06), pp 149–158. https://doi.org/10.1109/SSDBM.2006.27
12. Gray J, Liu DT, Nieto-Santisteban M, Szalay A, DeWitt DJ, Heber G (2005) Scientific data management in the coming decade. SIGMOD Rec 34(4):34–41. https://doi.org/10.1145/1107499.1107503
13. Guennebaud G, Jacob B et al (2010) Eigen v3. https://eigen.tuxfamily.org
14. Hinsen K (2013) Software development for reproducible research. Comput Sci Eng 15(4):60–63. https://doi.org/10.1109/MCSE.2013.91
15. Ishwarappa Anuradha J (2015) A brief introduction on big data 5Vs characteristics and Hadoop technology. Procedia Comput Sci 48:319–324. https://doi.org/10.1016/j.procs.2015.04.188
16. Könnecke M, Akeroyd FA, Bernstein HJ, Brewster AS, Campbell SI, Clausen B, Cottrell S, Hoffmann JU, Jemian PR, Männicke D, Osborn R, Peterson PF, Richter T, Suzuki J, Watts B, Wintersberger E, Wuttke J (2015) The NeXus data format. J Appl Crystallogr 48(1):301–305. https://doi.org/10.1107/S1600576714027575
17. Lofstead J, Ross R (2013) Insights for exascale IO APIs from building a petascale IO API. In: SC '13: proceedings of the international conference on high performance computing, networking, storage and analysis, pp 1–12. https://doi.org/10.1145/2503210.2503238
18. Nam B, Sussman A (2003) Improving access to multi-dimensional self-describing scientific datasets. In: CCGrid 2003. 3rd IEEE/ACM international symposium on cluster computing and the grid, 2003. Proceedings, pp 172–179. https://doi.org/10.1109/CCGRID.2003.1199366
19. Rew R, Davis G (1990) NetCDF: an interface for scientific data access. IEEE Comput Graph Appl 10(4):76–82. https://doi.org/10.1109/38.56302
20. Ritchie DM, Thompson K (1974) The UNIX time-sharing system. Commun ACM 17(7):365–375. https://doi.org/10.1145/361011.361061
21. Sanderson C, Curtin R (2016) Armadillo: a template-based C++ library for linear algebra. J Open Source Soft 1(2):26. https://doi.org/10.21105/joss.00026
22. Silvano C, Agosta G, Bartolini A, Beccari AR, Benini L, Besnard L, Bispo J, Cmar R, Cardoso JM, Cavazzoni C, Cesarini D, Cherubin S, Ficarelli F, Gadioli D, Golasowski M, Libri A, Martinovič J, Palermo G, Pinto P, Rohou E, Slaninová K, Vitali E (2019) The ANTAREX domain specific language for high performance computing. Microprocessors Microsyst 68:58–73. https://doi.org/10.1016/j.micpro.2019.05.005

23. Spreckelsen F, Rüchardt B, Lebert J, Luther S, Parlitz U, Schlemmer A (2020) Guidelines for a standardized filesystem layout for scientific data. Data 5(2). https://doi.org/10.3390/data5020043
24. Stroustrup B (2013) The C++ programming language, 4th edn. Addison-Wesley Professional
25. Sutton A, Holeman R, Maletic JI (2009) Abstracting the template instantiation relation in C++. In: 2009 IEEE international conference on software maintenance, pp 559–562. https://doi.org/10.1109/ICSM.2009.5306392
26. The HDF Group. Hierarchical data format, version 5 (1997-NNNN). http://www.hdfgroup.org/HDF5/
27. Touton J, Spertus M (2016) Declaring non-type template parameters with auto. http://www.open-std.org/jtc1/sc22/wg21/docs/papers/2016/p0127r2.html
28. Tratt L (2008) Domain specific language implementation via compile-time meta-programming. ACM Trans Program Lang Syst 30(6). https://doi.org/10.1145/1391956.1391958
29. Varga S: Easy to use HDF5 C++ templates for serial and parallel HDF5. https://github.com/steven-varga/h5cpp
30. Wells D, Greisen E, Harten R (1981) FITS—a flexible image transport system. Astron Astrophys Suppl 44:363
31. Wilkinson MD, Dumontier M, Aalbersberg IJ, Appleton G, Axton M, Baak A, Blomberg N, Boiten JW, da Silva Santos LB, Bourne PE et al (2016) The FAIR guiding principles for scientific data management and stewardship. Sci Data 3
32. Zhang W, Byna S, Niu C, Chen Y (2019) Exploring metadata search essentials for scientific data management. In: 2019 IEEE 26th international conference on high performance computing, data, and analytics (HiPC), pp 83–92
33. Zhang W, Byna S, Tang H, Williams B, Chen Y (2019) Miqs: metadata indexing and querying service for self-describing file formats. In: Proceedings of the international conference for high performance computing, networking, storage and analysis, SC '19. Association for Computing Machinery, New York, NY, USA. https://doi.org/10.1145/3295500.3356146

Kernelized Transfer Joint Matching for Unsupervised Domain Adaptation

A. K. Devika, Rakesh Kumar Sanodiya, and Babita R. Jose

1 Introduction

Transfer learning is referred to as a situation where what has been studied and learned in one scenario is adapted to different scenarios to improve generalization performance [1]. Traditional machine learning methods such as k-nearest neighbour (KNN) or support vector machine (SVM) are isolated methods where learning is purely based on a particular or sole task and a data set (i.e. training and test data must come from same task). This is a major drawback associated with these methods. However, the knowledge can also be transferred from one task to another if they are related.

In real world, the images or computer vision-related data cannot be always labelled. Figure 1 shows the images of a bicycle, a cup and a chair from two separate domains. First row images, for example, show the pictures belonging to source domain (A, B and C) of a particular distribution, and second row images belong to target domain(D, E and F)of another distribution. Even though images A and D are showing the pictures of a tea cup belonging to different distributions, the classifier is not able to correctly understand that labels of both the images are the same(that is, both the images are showing a tea cup). This is due to the training data that may contain a bias but is imperceptible to humans. So in any image classification problem, domain adaptation has become a necessity. That is, the classifier must correctly identify the target image for any change in domain. Specifically, domain adaptation is a problem in which the task space is the same and the difference is in input domain distributions.

A. K. Devika (✉) · B. R. Jose
Cochin University of Science and Technology, Kochi, Kerala, India
e-mail: devikaak@cusat.ac.in

R. K. Sanodiya
Indian Institute of Information Technology, Sri City, Chittoor, India
e-mail: rakesh.s@iiits.in

Fig. 1 Images of three objects(cup, bicycle and chair) from two different (source and target) domains

Traditional approaches have done the classification assuming the distributions to be same. But in cross-domain problems, the source and target distributions are usually different with different marginal probabilities. Thus, the major challenge is in lessening the distribution difference between source and target domains. Many research works in the related field have studied the subject in two different dimensions or categories like (1) sharing a common feature depiction that can reduce the dissimilarity between the source and target distributions and (2) reweighting the sample instances of source domain in order to reduce the gap between their respective distributions.

In our paper, we have jointly performed the feature mapping by decreasing the maximum mean discrepancy(MMD) distance in an infinite-dimensional reproductive kernel Hilbert space (RKHS) adapting the distributions of both the domains with reweighting source instances using Laplacian regularization.
The summary of our work is the following:

1. Present a novel technique called KTJM that aims to surmount the shortcomings that are present in other subsisting transfer learning approaches. This method has minimized marginal distributions mismatch of source and target domain incorporating maximum mean discrepancy metric using various popular kernels like Gaussian RBF, polynomial and linear kernels.
2. We have also included conditional distribution adaptation along with instance reweighting of source data that fills the gap of shortcomings of the previous approaches. This is then given to a Laplacian regularizer for further improving the performance.

2 Related Work

Here, we will discuss some of the existing works related to our proposed method. According to the literature survey, the transfer learning approaches can be categorized into four that includes instance-based transfer learning, parameter-based transfer learning, relational knowledge transfer learning and feature-based transfer learning. In the first approach, the instances or samples of the source are downweighted to match data sources and targets. In parameter-based transfer learning approaches, hyperparameters from the model form the learning factors to match source and the target task. In relational knowledge approach, it is assumed that source and target data have some similar relationship. In feature-based approach, a low-dimensional feature space is discovered and shared through nonlinear transformation such that the sources and targets are made closer [2].

Our work is based on feature-based transfer learning technique. Recent works on this area points to basically two categories of learning methods. First one is feature matching [3] that aims to decrease the change in marginal probability distribution between source and target domains. For example, Pan et al. [4] proposed a feature or dimension reduction method called maximum mean discrepancy embedding (MMDE) to find a latent space that can reduce the deviation of the source and target data distributions. This method fails to express due to the computation complexity of the expensive semi definite programming involved. TCA [5] is another method that finds good feature representation for domain adaptation. It tries to learn some transfer components across domains in an RKHS such that maximum mean discrepancy metric between both the distributions is lessened using a low rank representation. This representation helped in avoiding computational burden of MMDE. Standard classification (like kNN or SVM) or regression algorithms can be run on these transformed space to perform classification. The methods like TCA had taken only marginal distribution into consideration. Long et al. put forward a concept called joint distribution adaptation which jointly considers both marginal probability distribution and conditional probability distribution and constructed new feature domain representation for adaptation [6].

Including instance reweighting term in the problem statement can further improve the classification accuracy. The second category of methods emphasized on downweighting the source sample instances that are irrelevant. Transfer joint matching is such a method that downweights the source instances by implementing l2 norm sparse penalty on it [7]. But it did not consider the label space that is different during the real-time applications, and therefore, the accuracy was less for the conventional data sets. Joint geometrical and statistical alignment for visual adaptation(JGSA) [8] is a method proposed by Zhang et al.,which lessen the domain shift statistically and geometrically. It addresses various problems of the previous approaches and adds properties like maximizing target domain variance, preserving source domain discriminative information and distribution, subspace difference minimization, etc. Another approach called KUFDA was suggested by Sanodiya et al. [9]that showed a good improvement in the accuracies using VGG features. But still it did not consider the outlier samples much which reduced its accuracy. Optimal Transport for Domain Adaptation (OTGC) [10] is a popular technique that have used optimal transport theory for computing distances between probability distributions. To align the representations in the source and target domains, the work provides a regularized unsupervised optimum transportation model. Correlation alignment (CORAL) [11] is a method that aligns the second-order statistics of the source and target distributions with a linear transformation. This was further extended to learn nonlinear transformation through correlations of layer activations in deep neural network. Domain-invariant and class discriminative (DICD) representations [12] are a method that deliberately reduces the distance between each pair of projected samples with the same label. It also aims to maximize the distance between each pair of projected images from distinct classes. Distribution matching machine [13] is based on the structural risk minimization principle, which learns a transfer support vector machine by extracting invariant feature representations and estimating unbiased instance weights that jointly minimize the cross-domain distribution discrepancy. In an approach called JDDA-C (joint domain alignment and discriminative feature learning for unsupervised deep domain adaptation) [14], two-stream CNN architecture comprising of source and target data with shared weights is adopted. Here, an extra discriminative loss is proposed to encourage the shared representations to be more discriminative in nature. Manifold Embedded Distribution Alignment (MEDA) is a visual domain adaptation approach that uses structural risk minimization to learn a domain-invariant classifier in the Grassmann manifold [15]. But all these works have not explored the use of different types of kernels in the nonlinear feature space and many of them fails to preserve many of the objectives necessary to achieve domain adaptation effectively. Here, we have studied the performance of kernels like linear, RBF (Gaussian) and polynomial kernels in the proposed work exploiting different parameters like subspace dimension,tradeoff parameters, classifier parameter, etc.

3 Kernelized Transfer Joint Matching for Unsupervised Domain Adaptation

This section proposes a detail study on kernelized transfer feature learning-based approach through joint matching for domain adaptation(KTJM).

3.1 Problem Definition

In the real world, we do not have the perfect training set for the real classification problems. In order to study the real-life classification problems, mathematical formulation of the problem statement is necessary. Lets begin with the some of the definitions of terminologies. Given a training set $X = \{x^i\}_{i=1}^N$, there exists some probability distribution P(X) over our training set. Our label spaces Y include the possible labels of our samples in our problem where the label set is $Y = \{y^i\}_{i=1}^N$.

A learning domain D is defined as D={X,P(X)} and a learning task T is defined as T={$Y, P(Y|X)$}. A classifier function $f(x)$ with the probability distribution function $P(Y|X)$ predicts output variable or label for every new instance x. Furthermore, for a given source domain D_s with learning task T_s and unlabeled target domain D_t with learning task T_t, the goal of transfer learning is to use knowledge to increase learning of the target function in D_s and T_s when: $D_s \neq D_t$ or $T_s \neq T_t$. Given a labelled source domain $D_s = (x_1, y_1), (x_2, y_2), \ldots, (x_{n_s}, y_{n_s})$ and unlabelled target domain $(x_1, x_2, \ldots, x_{n_t})$ under the assumptions $X_s = X_t, Y_s = Y_t, P(X_s) \neq P(X_t), P(Y_s|X_s) \neq P(Y_t|X_t)$, the classifier must be able to reduce the difference in D_s and D_t.

In this study, we are going to propose a technique called KTJM that has good accuracy due to its properties like property preservation, distribution adaptation and instance reweighting. Let us study each and every component of the technique in detail.

3.2 Low-Dimensional Projection of Data

While performing a classification task, when the data is made to work on original feature space, the computation complexity always increases. Another disadvantage of high-dimensional data is overfitting. This is because the number of samples taken for the exact classification must be significantly greater than the square of the data dimension. And therefore, dimension reduction is a necessary technique to be adopted in any domain adaptation algorithm. Here, principal component analysis (PCA) is the dimensionality reduction approach used and it involves projecting the input data matrix X onto a set of d eigenvectors.

If $X = [x_1, x_2, x_3, \ldots x_n] \in R^{m \times n}$ is the input data matrix, $H = I_n - \frac{1}{n}1$ is the centering matrix, where I_n is the identity matrix of size $n = n_s + n_t$ and 1 is an $n \times n$ matrix of all ones , then the covariance or scatter of a matrix X is XHX^T. According to PCA, finding the orthogonal basis for input data is performed through a transformation matrix $G \in R^{m \times k}$ such that the variance in embedded data is maximized.

$$\max_{G^T G=I} \text{tr}(G^T X H X^T G) \tag{1}$$

The trace of the matrix is denoted by tr(.). This is an optimization problem that can be found out by eigen decomposition $XHX^T G = G\phi$, where $\phi = \text{diag}(\phi_1, \ldots, \phi_k) \in R^{k \times k}$ having d largest eigenvalues. Thus, the optimal d-dimensional representation is found out by using the equation $Q = [q_1, q_2, \ldots q_n] = G^T X$.

Kernelization: Kernels make the mathematical computation easier by incorporating kernel matrices. It is a tool that lets you map nonlinear data into a high-dimensional space by computing the inner products of all pairs of data in the feature space. In RKHS, let us consider a mapping τ that maps input data x into $\tau(x)$ such that $\tau(x) = [\tau(x_1), \tau(x_2), \ldots, \tau(x_n)]$ and kernel matrix $K = \tau(X)^T \tau(X) \in R^{n \times n}$. According to representer theorem, the PCA is kernelized using $G = \phi(X)B$ and is derived as

$$\max_{B^T B = I} \text{tr}(B^T K H K^T B) \tag{2}$$

where $B \in R^{n \times k}$ is the matrix of transformation for kernelization and the embedding after transformation becomes $Q = B^T K$. Choosing the right kernel is therefore a challenging task to achieve the desired goal. We investigate the performance of kernels such as linear, Gaussian RBF kernel and polynomial kernel in this paper.

$$\left. \begin{array}{l} \text{Linear Kernel: } K(x, y) = x^T y \\ \text{Gaussian RBF Kernel: } K(x_i, x_j) = exp(-\gamma \|x_i - x_j\|^2) \\ \text{Polynomial Kernel: } K(x_i, x_j) = (x_i . x_j + 1)^d \end{array} \right\} \tag{3}$$

3.3 Feature Matching Using Marginal Distribution

We know that the marginal probability distributions of source domain data and target domain data are always different in real-world circumstances,i.e. $P(x_t) \neq P(x_s)$. In order to reduce this distribution difference, it can be assumed that there is a feature map ϕ so that the distribution of the data that has been mapped becomes $P(\phi(x_s)) \approx P(\phi(x_t))$. Now, we find the distance between these distributions $P(\phi(x_s))$ and $P(\phi(x_t))$ using the metric maximum mean discrepancy in reproductive kernel Hilbert space(RKHS). If n_s and n_t represent the number of samples in source and target domain, MMD distance can be calculated as:

$$\left\| \frac{1}{n_s} \sum_{i=1}^{n_s} B^T K_i - \frac{1}{n_t} \sum_{j=n_s+1}^{n_s+n_t} B^T K_j \right\|_F^2 = tr(B^T K M K^T B) \tag{4}$$

where M is the MMD matrix and is computed as

$$M_{ij} = \begin{cases} 1/n_s n_s, & x_i x_j \in D_s \\ 1/n_t n_t, & x_i x_j \in D_t \\ -1/n_s n_t, & \text{otherwise} \end{cases} \tag{5}$$

Minimizing Eq. (4) and maximizing Eq. (2) are a possible solution to bring marginal distributions closer under the new representation $Q = B^T K$. Minimizing the distance can bring the data closer and maximizing the variance can make the features discriminative.

3.4 Transformation Using Conditional Distribution Adaptation

Reducing the marginal distribution difference alone cannot make conditional distribution difference less. There are various cases when the class balance is very much different in source and target distributions. In these type of situations, domain space will be the same but learning tasks are different. This gives rise to the idea of intraclass transfer.

We cannot directly model $P(y_t|x_t)$ because there is not any labelled data for the target domain. Thus, the basic idea is to create fictitious labels for the target data, and then, use intraclass correlation to adaptively reduce spatial dimension by applying some basic classifiers like KNN, SVM, etc. Here, we need to calculate maximum mean discrepancy of each class. Thus, class conditional probabilities become $P(Y_s|X_s = C)$ and $P(Y_t|X_t = C)$ with respect to class C rather than posterior probabilities $P(Y_s|X_s)$ and $P(Y_t|X_t)$.

The MMD between class conditional probability distributions $P(x_s|y_s = C)$ and $P(x_t|y_t = C)$ is

$$\left\| \frac{1}{n_s^{(c)}} \sum_{x_i \in D_s^{(c)}} B^T K_i - \frac{1}{n_t^{(c)}} \sum_{x_i \in D_t^{(c)}} B^T K_j \right\|_F^2 = tr(B^T K M_c K^T B) \qquad (6)$$

where K_i and K_j is the kernelized data samples of source and target domain, $D_s^{(c)}$ = $\{x_i : x_i \in D_s$ and $y(x_i) = C\}$, $D_s^{(c)}$ is the source data samples pertaining to class C, $y(x_i)$ is the true label of x_i and $n_s^{(c)}=|D_s^{(c)}|$. Similarly for target domain, $D_t^{(c)}$ =$\{x_j : x_j \in D_t$ and $y(x_j) = C\}$, $D_t^{(c)}$ is the target data samples pertaining to class C, $y(x_j)$ is the true label of x_j and $n_t^{(c)}=|D_t^c|$. MMD matrix M_c computed for each class is given by

$$M_{ij} = \begin{cases} \frac{1}{n_s^c n_s^c} & x_i, x_j \in D_s^c \\ \frac{1}{n_t^c n_t^c} & x_i, x_j \in D_t^c \\ \frac{-1}{n_s^c n_t^c} & \begin{cases} x_i \in D_s^c x_j \in D_t^c \\ x_j \in D_s^c x_i \in D_t^c \end{cases} \\ 0 & \text{otherwise} \end{cases} \qquad (7)$$

Equation (6) is minimized in this process such that Eq. (2) is maximized so that the conditional distributions are made closer using the representation $Q = B^T K$.

3.5 Downweighting Source Instances

To effectively perform domain adaptation and bring the source and target features closer, we have to correctly identify the instances of source that are closer to target domain and down weight the ones that are irrelevant. This is done through instance reweighting. In this paper, a l_{21} norm structured sparsity regularizer is imposed on transformation matrix B [7]. Each sample of B corresponds to an instance, and hence, sparsity in the rows facilitates distance reweighting adaptively. Thus, instance reweighting regularizer is introduced as

$$\|B_s\|_{2,1}^2 + \|B_t\|_F^2 \tag{8}$$

where $B_s := B_{1:n_s}$ is the matrix of transformation corresponding to samples or instances of source and $B_t := B_{n_s+1:n_s+n_t}$ is the transformation matrices according to target instances. So, here Equation (8) is to be minimized such that kernelized variance in Eq. (2) is maximized. This adaptive instance reweighting is observed in the new representation $Q = B^T K$.

3.6 Laplacian Regularizer

This is a method of regularization using Laplacian matrix. A given set of data points $\{x_1, x_2, \ldots x_n\}$ with pairwise similarities or distances e_{ij} can be transformed into a graph using a variety of methods. The purpose of developing a similarity graph is to model the local neighbourhood relationships of data points. A Laplacian matrix is a graph's matrix representation. We want to partition the graph so that edges between groups have relatively low weights (indicating points are dissimilar) and data in the same cluster has high weights. The pairwise affinity matrix can be defined as

$$W_{ij} = \begin{cases} e(x_i, x_j), & \text{if } x_i \in N_k(x_j) \mid x_j \in N_k(x_i) \\ 0, & \text{otherwise} \end{cases}$$

where $e(x_i, x_j)$ is a similarity or distance measure to measure the distance between the points. The paper has used cosine distance as the weight mode of the algorithm that determines the cosine distance between the two data samples(in order to determine their similarity). For grouping the data, the k-nearest neighbour (KNN) method is employed, with $N_k(x_j)$ designating the set of k-nearest neighbours of point x_i.

The Laplacian matrix L is given by $L = (D - W)$, where D is the degree matrix or diagonal matrix defined as $D_{ii} = \sum_j W_{ij}$, the sum of edge weights in the graph [15].

Thus, the regularizer term can be derived as:

$$R = \sum_{i,j=1}^{n} (Z(x_i) - Z(x_j))^2 W_{ij}$$

$$= \sum_{i,j=1}^{n} Z(x_i) L Z(x_j) \tag{9}$$

$$= \mathrm{tr}(B^T K L K^T B)$$

where tr(.) gives the trace of the matrix, K represents the kernelized input data samples and B is the transformation matrix.

3.7 Optimization of Objective Function

The overall objective function of this work can be obtained by combining all these essential components like distribution difference minimization, reweighting of source domain samples and Laplacian regularization. So by combining all the equations formulated above, we get

$$\min_{B^T K H K^T B = I} \mathrm{tr}\,(B^T K (M_c + \eta L) K^T B)) + \lambda(\|B_s\|_{2,1}^2 + \|B_t\|_F^2) \tag{10}$$

where λ is a regularization parameter that is used to balance instance downweighting and thereby match the features, η is Laplacian regularizer tradeoff parameter.

Optimization We derive the Lagrangian function in the optimization theory using Lagrange multiplier ϕ for which we denote $\phi = \mathrm{diag}(\phi_1, \phi_2, \dots \phi_k) \in R^{k \times k}$

$$L = \mathrm{tr}\,(B^T K (\sum_{c=0}^{C} M_c + \eta L) K^T B) + \lambda\|B_s\|_{2,1}^2 + \|B_t\|_F^2) + \mathrm{tr}\,(I - B^T K H K^T \phi) \tag{11}$$

Setting $\frac{\partial L}{\partial B} = 0$, we get generalized eigen decomposition problem as

$$(K (\sum_{c=0}^{C} M_c + \eta L) K^T + \lambda G) B = K H K^T B \phi \tag{12}$$

$\|B_s\|_{2,1}^2$ is a function which is not smooth at zero. Now its subgradient is computed as

$$\frac{\partial(\|B_s\|_{2,1}^2 + \|B_t\|_F^2)}{\partial z} = 2GB \tag{13}$$

where G is a diagonal matrix of subgradients with ith member equal to

$$
\begin{aligned}
G_{ii} &= \frac{1}{|a^i|}, k_i \in D_s, a^i \neq 0 \\
&= 0, k_i \in D_s, a^i = 0 \\
&= 1, k_i \in D_t
\end{aligned}
\tag{14}
$$

The above problem in Eq. (12) can be obtained by solving for d leading eigenvalues and eigenvectors in eigen decomposition problem. This reduces the adaptation matrix B, and thus, we obtain the optimal set of data vector which are discriminative in nature.

ALGORITHM: KTJM :Kernelized Transfer Joint Matching (KTJM) for Unsupervised Domain Adaptation

---

Input:Data X, subspace base dimension d , λ the regularization parameter for feature matching and reweighting the instances,k number of nearest neighbors in the KNN algorithm,η the tradeoff parameter to perform Laplacian regularization
Output: Output matrix Q_s and Q_t and Accuracy A
begin:

1. Find source and target domain input matrices X_s and X_t
2. Compute Kernel matrix K by Equation (3),centering matrix H and MMD matrix M using Equation (5). Initialise the gradient matrix G as an identity matrix.
3. Compute the laplacian Matrix L
   Repeat

   – Solve the eigen decomposition problem by Equation (12) and
   – Construct the adaptation matrix Q matrix using the equation $Q = B^T K$
   – Derive the embeddings Q_s and Q_t from the adaptation matrix
   – Train the KNN classifier on the embeddings to get pseudo target labels y_i
   – Construct MMD matrices M_c by (7)
   – Update the gradient matrix G by (14)

   until Convergence
4. Find the accuracy using $A = y_i/n_t$

---

4   Experiments

4.1   Setup

Office-Caltech and PIE data sets are two of the most extensively used benchmark data sets for analysing adaptation in the visual domain [16]. Four domains, namely C (Caltech), W (Webcam), D (DSLR)and A (Amazon) are included in Office-Caltech data set. In fact, Office-Caltech data set is constructed from 10 common classes of data sets, namely—Caltech-256 (which contains 256 classes of C) and Office-31 (which contains 31 classes of A, D and W). Each of the database consists of images taken from Amazon website and those taken with varying lighting conditions and position changes using a less quality camera (webcam) or a DSLR camera in different backgrounds. Decaf6 features (A Deep Convolutional Activation Feature for Generic Visual Recognition) are features extracted from trained deep CNN. They are named Decaf6 features since it is extracted from fully 6th connected layer of the CNN trained on ImageNet database. This gives more accuracy when compared to feature extractors like SURF. In the Office-Caltech data set, out of these 4 domains available, we choose two domains at random as the source and target domains and totally we get $4 * 3 = 12$ cross-domain mappings to perform experimentation. Pose, Illumination and Expression (PIE) data set of human faces was another data set used for further study of the work. This is a database of 40,000 facial images of 68 people. 13 different postures with 4 distinct people's expressions under various lighting situations are recorded and stored for creating this database. We have taken five different poses of people for study like C05(PIE1), C07(PIE2), C09(PIE3), C27(PIE4) and C29(PIE5). Taking a set of data for both source and target domains, we get $5 * 4 = 20$ different possible combinations of cross domain tasks.

4.2 t-SNE Feature Visualization

t-Distributed Stochastic Neighbour Embedding (t-SNE) is a method for visualizing data sets with multiple dimensions [17]. Figure 2a–d shows the t-SNE visualization of Office-Caltech10 Decaf6 features. Figure 2a shows the Decaf feature set of input features, where the source and target data are observed as non-classified features spread across the space, whereas Fig. 2b shows the classified output feature space of source and target data after performing classification using KTJM linear method. Figure 2c shows the source and target data output features classified after implementing KTJM RBF method, and Fig. 2d shows the output feature space after using KTJM polynomial method. Here, we used two unlike marks (plus and circle) to represent the source and target domain features. For effective visualization, ten distinct colours were employed to differentiate ten different classes. As shown in Fig. 2b–d the data features of both classes are observed to be matched using KTJM linear, RBF and polynomial methods, respectively.

The data points of each unique class are gathered and aggregated on a specific place after classification. Green plus('+') points and circle points, for example, are gathered together in a space at one location, black plus and circle points at another location, and ten different colour points denoting ten distinct classes are clustered at random locations, showing they are all classified. But as we can see in RBF and polynomial methods, the points are not clustered much effectively as compared to the linear method. This is because the classification accuracy of KTJM is found to be better in comparison with RBF and polynomial methods.

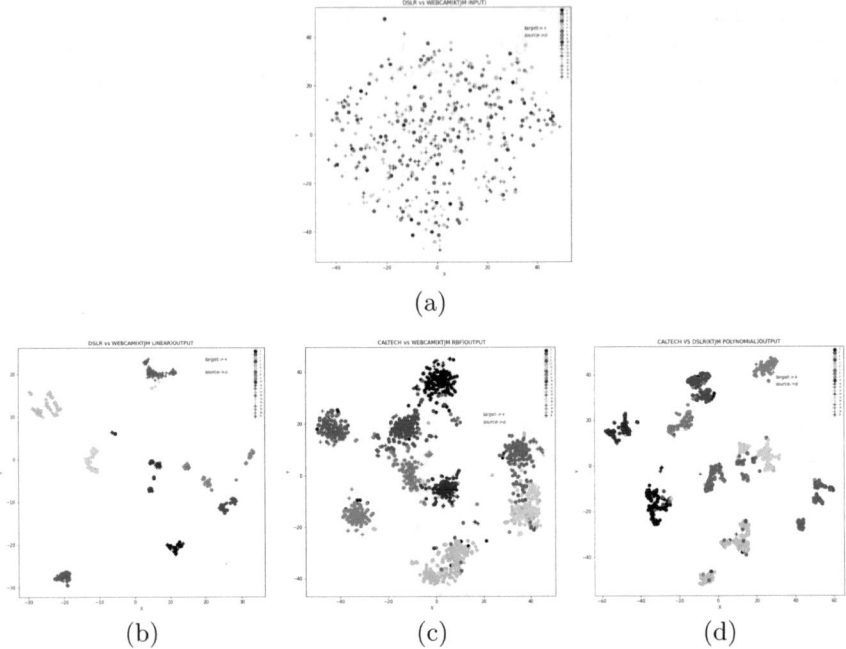

Fig. 2 a Represents input data visualization, **b** represents output data visualization using Linear kernel, **c** represents output data visualization using RBF kernel, **d** represents output data visualization using polynomial kernel.

5 Results and Discussion

Experiments were carried out on the various data sets described above. Tables 1 and 2 depict a comparison of the proposed work with various traditional transfer learning algorithms. The study was done varying different parameters of the algorithm. The parameters varied are subspace base dimension d that decides the dimension of the output matrix after adaptation, λ which is the regularization parameter for down-weighting and feature matching, k that denotes the number of nearest neighbours in the KNN algorithm to form Laplacian matrix and finally η which is the Laplacian regularization tradeoff parameter. The work was kernelized using different kernels like linear kernel, Gaussian kernel (RBF) and a polynomial kernel. While implementing RBF kernel, parameter γ which is the bandwidth of Gaussian kernel was varied to achieve maximum accuracy.

A thorough search was done on our work varying all these parameters. With an increment of 10, the subspace dimension d is adjusted for different values ranging from 80 to 180, and a dimension value of 100 was found to be optimum for PIE data set as shown in Fig. 3b. The trade off parameter λ was varied from 10^{-4} to 10^2, and highest accuracy was achieved for the value $\lambda = 0.1$. Then, we adjusted the η

Table 1 Accuracies (%) on the PIE face data set.

Methods	Data sets																				Average
	PIE 1	PIE 2	PIE 3	PIE 4	PIE 5	PIE 6	PIE 7	PIE 8	PIE 9	PIE 10	PIE 11	PIE 12	PIE 13	PIE 14	PIE 15	PIE 16	PIE 17	PIE 18	PIE 19	PIE 20	
	5→7	5→9	5→27	5→29	7→5	7→9	7→27	7→29	9→5	9→7	9→27	9→29	27→5	27→7	27→9	27→29	29→5	29→7	29→9	29→27	
PCA	24.8	25.18	29.26	16.3	24.22	45.53	53.35	25.43	20.95	40.45	46.14	25.31	31.96	60.96	72.18	35.11	18.85	23.39	27.21	30.34	33.85
kNN	26.08	26.59	30.67	16.67	24.49	46.63	54.07	26.53	21.37	41.01	46.53	26.23	32.95	62.68	73.22	37.19	18.49	24.19	28.31	31.24	34.76
JDA	58.81	54.23	84.5	49.75	57.62	62.93	75.82	39.89	50.96	57.95	68.45	39.95	80.58	82.63	87.25	54.66	46.46	42.05	53.31	57.01	60.24
TCA	40.76	41.79	59.63	29.35	41.81	51.47	64.73	33.7	34.69	47.7	56.23	33.15	55.64	67.83	75.86	40.26	26.98	29.9	29.9	33.64	44.75
DICD	73	72	92.2	66.9	69.9	65.9	85.3	48.7	69.4	65.4	83.4	61.4	93.1	90.1	89	75.6	62.9	57	65.9	74.8	73.1
TJM	29.52	33.76	59.2	26.96	39.4	37.74	49.8	17.09	37.39	35.29	44.03	17.03	59.51	60.58	64.88	25.06	32.86	22.89	22.24	30.72	37.29
LDADA	34.5	44.9	61.5	35.4	31.4	34.9	53.5	26.4	38.2	30.5	60.6	40.7	61.3	56.7	67.8	50.4	31.3	24.1	35.4	48.2	43.4
JGSA	68.07	67.52	82.87	46.5	25.21	54.77	58.96	35.41	22.81	44.19	56.86	41.36	72.14	88.27	86.09	74.32	17.52	41.06	49.2	34.75	53.39
MEDA	49.6	48.4	77.23	39.82	58.49	55.27	81.25	44.05	56.24	57.82	76.23	53.06	85.95	78.2	80.2	67.70	57.71	49.66	62.13	72.15	62.56
KUFDA	67.67	70.34	86.06	49.02	72.62	74.34	87.86	61.7	73.91	72.56	86.96	69.85	90	88.4	84.62	75.24	54.05	67.46	70.77	76.78	74.42
KTJM LINEAR	**73.66**	**74.63**	**94.11**	**60.85**	**70.26**	**80.09**	**89.76**	**64.83**	**81.9**	**79.86**	**94.17**	**72.37**	**94.78**	**94.9**	**94.06**	**81.19**	**63.6**	**67.53**	**72.55**	**81.74**	**79.342**
KTJM RBF	**61.76**	**62.99**	**86.51**	**58.33**	**62.06**	**68.14**	**81.14**	**50.12**	**69.42**	**66.61**	**86.72**	**59.44**	**89.41**	**87.54**	**90.26**	**71.08**	**55.79**	**49.85**	**58.95**	**68.61**	**69.2**
KTJM POLYNOMIAL	**59.67**	**67.03**	**90.12**	**56.31**	**70.98**	**74.82**	**87.35**	**60.05**	**76.41**	**74.03**	**90.48**	**69.67**	**91.87**	**92.2**	**91.91**	**78.06**	**64.41**	**57.46**	**67.65**	**75.22**	**74.78**

Table 2 Accuracies (%) on the Office-Caltech10 data set

Methods	Data sets												Average
	Caltech versus Amazon	Caltech versus webcam	Caltech versus Dslr	Amazon versus Caltech	Amazon versus webcam	Amazon versus Dslr	webcam versus Caltech	webcam versus Amazon	webcam versus Dslr	Dslr versus Caltech	Dslr versus Amazon	Dslr versus webcam	
JDA	90.19	85.42	85.99	81.92	80.68	81.53	81.21	90.71	100	80.32	91.96	99.32	87.44
OTGL	92.15	84.17	87.25	85.51	83.05	85	81.45	90.62	96.25	84.11	92.31	96.29	88.18
JGSA RBF	91.13	83.39	92.36	84.86	80	84.71	84.51	91.34	100	84.77	91.96	98.64	88.97
JGSA LINEAR	91.75	85.08	92.36	85.04	84.75	85.35	84.68	91.44	100	85.75	92.28	98.64	89.76
KTJM LINEAR	**91.86**	**82.71**	**92.36**	**86.73**	**82.37**	**88.54**	**87**	**91.34**	**100**	**87.44**	**92.48**	**99.66**	**90.2**
KTJM RBF	**92.07**	**85.76**	**86.62**	**86.55**	**77.29**	**85.35**	**85.93**	**90.4**	**100**	**87.36**	**92.9**	**99.66**	**89.15**
KTJM POLYNOMIAL	**89.77**	**75.93**	**80.89**	**82.19**	**66.78**	**85.35**	**66.25**	**72.03**	**100**	**83.7**	**87.58**	**100**	**82.53**

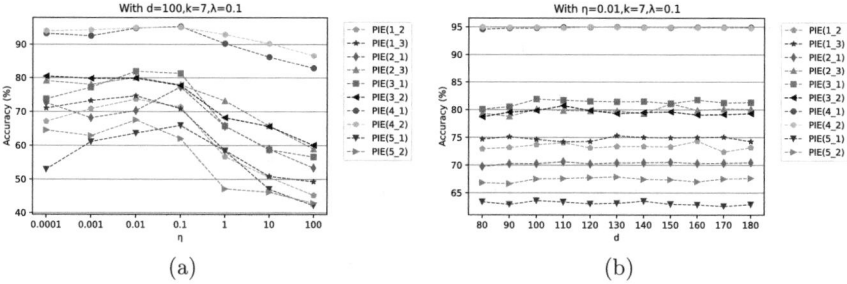

Fig. 3 **a** Accuracy (%) versus η, **b** accuracy (%) versus d

parameter values from 10^{-4} to 10^2, as illustrated in Fig. 3a. The number of nearest neighbours in the KNN algorithm k in forming a Laplacian matrix was the other parameter to investigate. The optimal values of these parameters for acquiring this accuracy are $\eta = 0.01$ and $k = 7$. The optimal values of each and every parameter were found out by keeping all the other parameter values constant.

When the proposed approach was applied to the PIE data set with these parameters, an average accuracy of 79.34% was found for a KTJM linear method, 69.2% accuracy for KTJM RBF method and 74.78% accuracy for KTJM polynomial method, respectively. KTJM RBF has adapted a γ value of 0.6 and KTJM polynomial method has used a polynomial degree of 2 for obtaining this accuracy. KTJM is compared with all the primitive approaches like nearest neighbour, principal component analysis and recent existing domain adaptation methods like TCA [5], JDA [6], TJM [7], JGSA [8], DICD [12], MEDA [15] and KUFDA [9]. KTJM linear achieves a considerable performance improvement of 4.922% above the nearest approach KUFDA on the 20 cross-domain mappings, with an average classification accuracy above 75%. The basic methods TCA, TJM and JDA perform poorly in this data set with an average classification accuracy of 44.75%, 60.24% and 37.29%, respectively.

The proposed work KTJM was also run on the other benchmark data set Office-Caltech 10. The subspace dimension was adjusted from 10 to 60 in this experiment. This is due to the fact that the subspace dimension must be at least as large as the number of classes in the data set. Because the Office-Caltech data set includes ten classes and the PIE data set has 60, the dimension value range was chosen accordingly. The optimal parameters found for this data set are $d = 30$, $\lambda = 0.1$, $\eta = 0.1$ and $k = 5$. γ value of 0.6 is adopted for KTJM RBF method and a polynomial degree of 2 is found to achieve good accuracy for KTJM polynomial method. The proposed technique achieved average accuracy of 90.2% for the KTJM linear method, 89.15 % for the KTJM RBF method and 82.53 % for the KTJM polynomial method. On the 12 data sets, KTJM outperforms the JGSA approach by an increase of 0.44%, baseline approach JDA by 2.76% and OTGL by 2.02%. Out of the different kernelized proposed methods, KTJM linear achieved the maximum accuracy when compared to KTJM RBF and KTJM polynomial methods, and therefore, the linear kernel was found to perform better when compared to other kernel approaches for both Office-Caltech and PIE data sets.

6 Conclusion

In this novel method, we have projected the components in a transformed feature space by reducing the distance between both domains with a Laplacian regularizer. It has also reweighted the irrelevant instances across the domains. The proposed method KTJM works well with linear kernel among both the data sets and has achieved a good classification accuracy. This shows an effective improvement when compared to many of the existing domain adaptation strategies. This is due to the inclusion of the shortcomings found in other primitive transfer learning approaches. In future, the work can be further extended and studied in multi-source domain adaptation areas. The input data feature set extracted can also be changed to CNN extracted features like VGGNet features, ResNet features, etc.

References

1. Goodfellow I, Bengio Y, Courville A (2016) Deep learning, p 800
2. Zhuang F et al (2021) A comprehensive survey on transfer learning. Proc IEEE 109(1), 43–76
3. Sanodiya RK, Mathew J (2019) A framework for semi-supervised metric transfer learning on manifolds. Knowl Based Syst 176:1–14
4. Pan SJ, Kwok JT, Yang Q (2008) Transfer learning via dimensionality reduction. In: Proceedings of 23rd AAAI conference on artificial intelligence, Chicago, IL, pp 677–682
5. Pan SJ, Kwok JT, Yang Q (2011) Domain adaptation via transfer component analysis. IEEE Trans. Neural Networks 199–210
6. Long M, Wang J, Ding G, Sun J, Yu PS (2013) Transfer feature learning with joint distribution adaptation. In: Proceedings of the IEEE international conference on computer vision, pp 2200–2207
7. Long M, Wang J, Ding G, Sun J, Yu PS (2014) Transfer joint matching for unsupervised domain adaptation. In: Proceedings of the IEEE conference on computer vision and pattern recognition, pp 1410–1417
8. Zhang J, Li W, Ogunbona P (2017) Joint geometrical and statistical alignment for visual domain adaptation. In: CVPR
9. Sanodiya RK, Mathew J, Paul B, Jose BA (2019) A kernelized unified framework for domain adaptation. IEEE Access 7:181381–181395
10. Courty N, Flamary R, Tuia D, Rakotomamonjy A (2016) Optimal transport for domain adaptation. IEEE Trans Pattern Anal Mach Intell 99:1-1
11. Sun B, Feng J, Saenko K (2016) Return of frustratingly easy domain adaptation. In: AAAI, vol 6.8
12. Li S, Song S, Huang G, Ding Z, Wu C (2018) Domain invariant and class discriminative feature learning for visual domain adaptation. IEEE Trans Image Process 27:4260–4273
13. Cao Y, Long M, Wang J (2018) Unsupervised domain adaptation with distribution matching machines. In: Proceedings of the 2018 AAAI international conference on artificial intelligence
14. Chen C, Chen Z, Jiang B, Jin X (2019) Joint domain alignment and discriminative feature learning for unsupervised deep domain adaptation. In: Proceedings of AAAI conference on artificial intelligence, pp 3296–3303
15. Belkin M, Niyogi P, Sindhwani V (2006) Manifold regularization: a geometric framework for learning from labeled and unlabeled examples. J Mach Learn Res 7:2399-2434
16. Sanodiya RK, Mathew J, Saha S, Thalakottur MD (2019) A new transfer learning algorithm in semi-supervised setting. IEEE Access 7:42956–42967

17. Maaten LVD, Hinton GJ (2008) Visualizing data using t-SNE. Mach Learn Res 9:2579–2605
18. Fernando B, Habrard A, Sebban M, Tuytelaars T (2013) Unsupervised visual domain adaptation using subspace alignment. In: Proceedings of the IEEE international conference on computer vision, pp 2960–2967

Printed in the United States
by Baker & Taylor Publisher Services